Owls

Marianne Taylor

B L O O M S B U R Y
LONDON · NEW DELHI · NEW YORK · SYDNEY

Bloomsbury Natural History
An imprint of Bloomsbury Publishing Plc

50 Bedford Square　　　　1385 Broadway
London　　　　　　　　　　New York
WC1B 3DP　　　　　　　　　NY 10018
UK　　　　　　　　　　　　　USA

www.bloomsbury.com

BLOOMSBURY and the Diana logo are trademarks of Bloomsbury Publishing Plc

First published 2017

British Library Cataloging-in-Publication Data
A catalogue record for this book is available from the British Library.

Library of Congress Cataloguing-in-Publication data has been applied for.

ISBN:　　PB:　978-1-4729-3369-0
　　　　ePub:　978-1-4729-3371-3
　　　　ePDF:　978-1-4729-3370-6

2 4 6 8 10 9 7 5 3 1

Design by Rod Teasdale
RR Donnelley Asia Printng Solutions Limited

MIX
Paper from
responsible sources
FSC® C101537

To find out more about our authors and books visit www.bloomsbury.com. Here you will find extracts,
author interviews, details of forthcoming events and the option to sign up for our newsletters.

giving
nature
a home .

Contents

Meet the Owls

A haunting, fluted hoot that jolts you out of sleep at 4am. A white ghost freeze-framed in your headlights on a dusk drive home in winter. A furious little face staring at you from the crack in an old barn roof. Encounters with owls in Britain are often startling and always memorable. Few other birds evoke such a strong sense of magic and mystery and – if you meet their gaze – understanding and connection, or at least the illusion of such. It's almost impossible to look at the face of an owl and *not* attribute some human quality to its expression.

Secret lives

Wherever we live on the planet, almost all of us know an owl when we see one; its forward-facing eyes seem to burn into our own. Centuries of fanciful folklore have bumped owls higher up in our cultural consciousness than perhaps any other birds, even though most of us only see owls infrequently in our everyday lives, if at all. The nocturnal habits and deep forest habitats of some owls, coupled with predatory powers that seem to almost defy nature and a repertoire of haunting and frightening calls, endow them with a mystique as irresistible as it is potent.

Above: A Tawny Owl sitting at its tree-hole nest is the very image of a 'wise old owl' of the woods.

Opposite: The ethereal beauty of the Snowy Owl.

In this age of television, YouTube and Facebook, we now have opportunities to see owls in action in different settings, without having to take a midnight walk in the woods. Through skilled camerawork we can see for ourselves how owls use their remarkable senses to find

Above: Short-eared Owls seem to be more sociable than most other species, but this behaviour is still not well understood.

and catch their prey, overcoming challenging conditions that most other animals would find impossible to handle. We can also watch rescued or trained captive-bred birds interacting with their keepers, revealing depths of charm, adventurousness and intelligence that vastly add to their appeal. Yet the day-to-day lives of wild owls remain rather mysterious, though our knowledge and understanding of them is growing by the day.

In Britain, there are five species of owls regularly breeding in any numbers. That's a tiny proportion of the world total of more than 200 species. However, these five owls are quite a diverse bunch, in size, shape, colour, habits and personality. In short, they demonstrate five very different ways to be an owl. In addition to these five, a few other species (again, all very different) make occasional 'guest appearances' in the wild in Britain.

File under 'O'

The term 'bird of prey' denotes a bird that hunts and eats other vertebrate animals. Many birds, from rails to herons and gamebirds to songbirds, may do this from time to time. However, the title of 'bird of prey' usually applies only to the members of three distinct groups, all noted for their predatory ways: the hawks and their allies (eagles, kites, buzzards, harriers and the like); the falcons; and the owls. The differences between owls and the rest have long been acknowledged, and their distinction recognised. Traditionally hawks and falcons are known as 'diurnal birds of prey', or 'raptors'/'diurnal raptors', and in field guides they are treated separately to the owls.

Today, biologists can work at the genetic level to investigate how closely related various bird groups really are, and their research has found evidence that the owls may actually be more closely related to the hawks than the falcons are. In any case, the similarities between a buzzard, a kestrel and an owl owe more to convergent evolution (whereby animals that pursue similar lifestyles evolve similar characteristics over time – as with dolphins and fish, for example) than to relatedness. The bird-of-prey lifestyle has arisen separately in two or more distinct bird lineages. Being a hunter of other vertebrates is a good way for a bird to make a living if it is done right. It is much more difficult to catch a relatively large, clever animal than it is to pick up an insect, but just one good-sized 'kill' a day is often enough to

meet the hunter's nutritional needs. The specialist hunters in the bird world all have superlative senses to find active, intelligent prey, the brainpower to outwit it, the speed and strength to overcome and catch it, and the physical equipment of talons and a hooked bill to kill and consume it.

According to some systems of classification, the owls' closest relatives are not other birds of prey, but a group of insect-eaters, the nightjars. Although nightjars are not fearsome killers (unless you happen to be a moth), they do have traits in common with owls – a nocturnal lifestyle with all the sensory adaptations that entails, and beautifully camouflaged plumage so they can sleep all day without easily being discovered.

Above: The Goshawk, a typical example of the largest family of birds of prey.

Left: Like owls, nightjars are nocturnal, and chase down insect prey in flight.

Families and genera

Birds are grouped in the taxonomic class Aves, which is subdivided into about 30 major groups or orders. One of these is Strigiformes, the owl order, and it contains two families – Tytonidae (the barn owls and bay owls – about 16 species in all) and Strigidae (all the rest – nearly 200 species). The barn owls and bay owls are clearly distinct from other owls – they are relatively small-eyed and long-faced, with softly mottled (rather than more strongly spotted and barred) plumage patterns and long wings and legs. The barn-owls form the genus *Tyto*, the bay-owls the genus *Phodilus*.

The family Strigidae contains about 28 genera. The five largest genera, between them holding about three-quarters of all owl species, are *Otus* (the scops owls), *Ninox* (the boobooks or hawk-owls), *Glaucidium* (the pygmy owls), *Strix* (the wood owls), and *Bubo* (the eagle owls and fish-owls). The smaller genera, with only a handful of species apiece, include Asio, the eared owls, and *Athene*, the little owls. However, in terms of global distribution they punch well above their weight, with some very widespread species.

Each genus has its own set of distinctive traits. The scops-owls are very small, well-camouflaged owls with slim bodies and obvious ear tufts, and are only found in the Old World. The screech-owls are their New World equivalents. The *Ninox* owls are native to East Asia and Australasia and are the least 'owl-like' of owls, having relatively small heads and long tails, and a more hawk-like general appearance, with no ear tufts. Pygmy-owls also lack ear tufts. They are found mainly in the Americas and are very small, but are fierce predators for their size. The eagle-owls are the largest, and include some huge and very powerful predators. They have ear tufts and very bulky physiques. The wood-owls are medium to large woodland owls with excellent camouflage and particularly sedentary habits. Little owls are small and often diurnal owls of more open habitats, and the eared owls are mostly long-winged, very active hunters, and include some strongly migratory and widespread species.

Right: Sooty Owls are relatives of our Barn Owl, and live in forests in Australia.

Left: The bay-owls are a distinctive group within the family Tytonidae, with squat bodies and very elongated faces.

Above: White-fronted Scops-owl, an Asian member of the large scops-owl genus *Otus*.

Above: The genus *Ninox* includes Australia's Southern Boobook.

Above: The tiny pygmy-owls make up the genus *Glaucidium*.

Above: Screech-owls, genus *Megascops*, replace the scops-owls in the Americas.

Above: The largest of the wood-owls (genus *Strix*) is the Great Grey Owl.

Above: Barred Eagle-owl, a handsome south-east Asian *Bubo* species.

Above: The Burrowing Owl is a widespread American species of the genus *Athene*.

Above: The small genus *Asio* includes the Striped Owl of South America.

The Brits

Above: The beautiful and distinctive Barn Owl is Britain's only representative of the family Tytonidae.

Below: Most woodlands in Great Britain hold at least one pair of Tawny Owls.

Below right: The introduced Little Owl is most common in southern England.

The five owls that breed widely in Britain are the Barn Owl (*Tyto alba*), Tawny Owl (*Strix aluco*), Short-eared Owl (*Asio flammeus*), Long-eared Owl (*Asio otus*) and Little Owl (*Athene noctua*). All five species have quite an extensive world distribution – the Barn Owl, Long-eared Owl and Short-eared Owl in particular, with all three occurring in the Americas as well as Eurasia. The Tawny Owl is found over much of Eurasia. The same goes for the Little Owl – but this species is not native to Britain; it was introduced here in the late 19th century, and has also been introduced to New Zealand. Neither Tawny nor Little Owls are found in Ireland.

The Tawny is Britain's most numerous owl, with 50,000 breeding pairs. It is found over most of Great Britain, though is missing from the most upland parts of Scotland and is also absent from many island groups, including Shetland, Orkney and the Western Isles. The Little Owl is represented by 5,700 breeding pairs, mostly in England, and just reaching southern Scotland, but is scarce in Wales and the

far south-west of England. The Barn Owl is widespread in lowland Britain and Ireland and we have about 4,000 pairs. The Long-eared Owl breeds very patchily across the British Isles with 3,500 pairs, but becomes more numerous and widespread in winter as visitors from the Continent arrive in varying numbers. It is a similar story with the Short-eared Owl, with just 1,400 pairs, nearly all of them in Scotland and northern England, but in some winters up to 50,000 more birds may visit from eastern Europe.

We think of owls as forest birds, but only two of our five owls are really birds of woodland. The Tawny is found in deciduous woodland, wooded parkland and gardens, while the Long-eared is more likely in coniferous woodland adjoining fields or moorland, and more open lowland areas in winter (with stands of dense scrub for roosting). In Ireland, where there are no Tawnies, Long-eareds occupy all kinds of woodland.

The other three species are birds of more open country. The Barn Owl is a classic species of low-lying farmland with rough grassland and old farm buildings in which to nest, but it is also found on saltmarsh and lowland heath. The Short-eared Owl nests on moorland, and in winter is found on all kinds of open countryside but especially coastal grazing marsh. The Little Owl is another farmland bird but also occurs in parks and large rural gardens.

Below left: The Short-eared Owl is an early riser by owl standards, usually active from mid afternoon.

Below: The Long-eared Owl is our shyest owl, and the most difficult to observe.

Left:
A beautiful and boldly marked bird, the Northern Hawk-owl is active by day and easy to see.

Left:
Tengmalm's Owl is a small but fierce, strictly nocturnal forest species.

Left:
Male European Scops-owls sing incessantly through the night until they attract a mate.

Left:
Unmistakeable and spectacular, the Snowy Owl is a rare visitor from the high Arctic.

Special guests

In addition to these five, a few other owls appear on the official British bird list. The Snowy Owl (*Bubo scandiacus*), the familiar large white owl immortalised as 'Hedwig' in the Harry Potter films, is an Arctic species and a very rare winter visitor. It has also nested in Britain, most recently on Fetlar in the Shetlands, where one pair bred for several years in the late 20th century. Two other Arctic owls have reached Britain in winter a handful of times. The Northern Hawk-owl (*Surnia ulula* – not related to the *Ninox* hawk-owls) and Tengmalm's or Boreal Owl (*Aegolius funereus*) are both small owls which sometimes make 'irruptive' movements south-west from Scandinavia and Russia in winter, when weather or food shortages are making life difficult.

By contrast, the European Scops-owl (*Otus scops*), a tiny slim owl with a tufted head, is a very rare summer visitor. A long-distance migrant that winters in Africa and breeds in southern Europe, it occasionally 'overshoots' its breeding grounds and ends up in northern Europe. If a female European Scops-owl were to turn up in Britain, we would probably never know she was there, but males give themselves away by calling constantly, all night long, in a bid to attract a mate.

The final owl on the British list is a figure of controversy. The Eurasian Eagle-owl (*Bubo bubo*) is the world's biggest and heaviest owl species and is widely distributed across Europe and Asia. It is regularly seen in Britain and there are several recent breeding records, but the origin of these birds is hotly disputed. Without doubt, at least some of them are escaped falconry birds – this uber-owl is very popular with falconers. However, there is an argument that at least some may be wild birds that crossed over from the near Continent. Eurasian Eagle-owls are especially controversial because they can and do kill many other birds of prey, including other owls, and the area where they have bred in Britain (in the Bowland moors, Lancashire) is home to one of England's rarest breeding birds of prey, the Hen Harrier. Some who consider this owl definitely

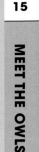

Above and right: Eurasian Eagle-owls are huge, powerful predators. They are sometimes mistaken for Long-eared Owls with their orange eyes and ear tufts, but are much bulkier.

non-native believe it is an invasive threat to native wildlife, and should be culled in Bowland and anywhere else it sets up home, to prevent it becoming established. Supporters of the wild-origin view believe that it should have all the same protections as other wild birds in Britain.

Some other owls, which are most definitely of captive origin, may be seen at large in Britain on occasion. The most likely include the Bengal Eagle-owl (*Bubo bengalensis*), Great Grey Owl (*Strix nebulosa*) and Great Horned Owl (*Bubo virginianus*). Native species, especially Barn Owls, are also sometimes encountered as escapees. Escapees are likely to be wearing jesses (leather straps on their legs, to which leashes can be attached) and may approach closely when they get hungry, as they associate people with free food. If you find an escaped owl, you can report it at the Independent Bird Register – independentbirdregister.co.uk

Above: A captive Bengal Eagle-owl wearing jesses. This is one of the most popular owl species used in falconry.

Anatomy and Adaptations

Across the world of birds, there is not a great deal of variation in general 'body plan'. All birds have two legs, two wings, feathers and a bill, and the majority of them get about by flying and walking. Compare that to the mammals, which are half as diverse in number of species but much more varied in structure with the likes of whales, bats, moles, giraffes, platypuses, kangaroos, lemurs, flying squirrels and humans, and it's clear to see how versatile and practical the basic bird body plan is. Owls are unmistakably birds, but equally they are not confusable with any other birds. Evolution has endowed them with many obvious and unique anatomical quirks that help them pursue their way of life.

Body basics

Owl anatomy is basically the same as that of the overwhelming majority of other birds. The foundation is a lightweight skeleton, optimised for flight, with some bones being hollow with supporting internal struts to save weight. The feet bear four clawed toes, and in owls the toe arrangement is zygodactyl – two toes point forwards and two backwards. The 'hand' bones are elongated and simplified, with only the thumb being a separately moveable digit (forming the alula, which can be extended to generate extra lift at low flight speeds) – the rest of the hand is a single structure that supports the flight feathers. The shoulder joints have exceptional mobility. The sternum is deep and keel-shaped for the attachment of powerful flight-propelling pectoral muscles. As another way to save weight, birds lack teeth; instead the bones of the jaw have a covering of horny but lightweight keratin, forming the bill. Although some birds have long tails, these are made of feathers only, and the tail bone is reduced to a short projection called a pygostyle.

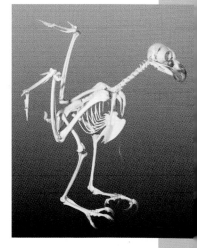

Above: The owl skeleton reveals how spindly these birds' bodies are beneath their dense coat of feathers.

Opposite: The staring yellow eyes set in a wide facial disk give this Short-eared Owl an imposing look.

Having wings for forelimbs, birds cannot manipulate objects the same way mammals do. However, they have very long, flexible necks with more neck bones than most other vertebrates, freeing them to use their bills for the sorts of tasks that mammals might carry out with their forefeet/hands. If you were to examine an owl skeleton, the length of the neck would probably be one of the things you noticed first, as would the forward projection of the strongly hooked bill. Owls look short-necked and flat-faced in life but their outline is shaped far more by feathers than by flesh and bone.

As well as lungs, birds have a system of internal interlinked air sacs which allow for a more efficient breathing cycle, such that the intake of fresh air is never interrupted. Contraction of the pectoral (chest) muscles during flight helps to push air around the system and drive the action of respiration.

Many birds, including owls, have a gizzard – an organ along their digestive tract that carries out the sort of heavy-duty grinding of food that would be done by chewing in animals with teeth. The gizzard is where pellets of indigestible food parts are formed – this allows owls to swallow prey whole or in large chunks without worrying about bones, hair or the tough bits of insect exoskeletons, as these will be regurgitated in pellet

Above: All owls cough up pellets of indigestible prey remains. Those of Snowy Owl could contain large intact bird and mammal skulls.

form some time after the meal. Birds also have very efficient kidneys which reclaim maximum fluid from food consumed, meaning that their urine is concentrated to a near solid (the white part in bird droppings). Lacking a bladder for storing excess fluid is yet another flight-friendly, weight-saving feature.

Senses

To hunt active, alert and intelligent prey, you need to have top-notch sensory apparatus. In the case of owls, vision and hearing are the most vital senses, hearing in particular. The distinctive shape of the owl head maximises the functionality of the eyes and ears. Smell and taste do not seem to be important or highly developed in owls, but they do rely on their well-developed sense of touch in certain circumstances.

Below: Harriers, like this female Hen Harrier, have rather flattened owl-like faces, and hunt in a very owl-like manner.

Eyes

These are certainly the first things we notice about owls, and having eyes that point forwards is the mark of a predator. Most other birds have eyes placed on the sides of their heads, providing an all-round field of view, but with relatively little overlap between their visual fields (in a few cases none at all), which means their binocular vision is inferior to that of an owl.

Having binocular vision (overlapping fields of view) means better depth perception – the ability to judge exactly how far away something is. This is achieved by the brain comparing the images received by each eye in the area of overlap.

Eye placement is inevitably a compromise between having a wide field of view and having good depth perception. For a prey species, it's vital that you see danger coming from any direction – the earlier you spot it, the better your chances are of getting away in time, so side-mounted eyes are usually best. Predators,

Eye colour

Above: Long-eared Owl.

Two of our owls (Barn and Tawny) have very dark brown, almost black eyes. Two more (Short-eared and Little) have bright yellow eyes, giving their faces a very intense look. The fifth, the Long-eared Owl, has beautiful rich orange eyes – though Long-eareds in North America are different, having paler, yellower eyes. Look at owls elsewhere in the world and you'll find some other colours – mid-brown, greenish-yellow – though it must be said that most species have very dark or yellow eyes. There is no obvious guiding factor behind owl eye colour. Among the British five, the yellow-eyed twosome are the most diurnal

Above: Tawny Owl.

Above: Barn Owl.

though, need good depth perception for accurately 'locking on' to a moving target, so tend to have flatter faces and more forward-facing eyes. This is obvious when we look at mammals – compare the face shape and eye placement of a deer or Rabbit to that of a Fox or a cat. Because birds fly through three-dimensional space, though, a broad field of view is generally more important to them than it is to earthbound mammals. The raptors that are active by day have more forward-facing eyes than most other birds, but among them only the harriers come close to the owl facial layout.

species, but this doesn't hold elsewhere in the world. While most of the more diurnal owls do have yellow rather than dark eyes, there are many more yellow-eyed owls that are strictly nocturnal. If there is something definite about owl biology or behaviour to be inferred by eye colour, we haven't yet discovered what it is.

Owls and other birds have a 'nictitating membrane' or third eyelid that flicks across the eye periodically to clean it, and also helps protect the eyes when the owl is dealing with prey. When caught in photos, this can give the eye a cloudy or milky appearance.

Above: Tawny Owl showing its *nictitating* membrane.

Above: Short-eared Owl.

Above: Little Owl.

Right: Owl eyes, like those of many nocturnal animals, have a reflective membrane behind the retina that produces 'eyeshine' when caught in the light.

Below: A flexible neck helps to make up for their immobile eyes.

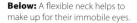

Most owls are nocturnal, so need eyes that work well in low light. The best way to improve the performance of an eye in near darkness is to give it a very big retina – the light-gathering component at the back of the eyeball. For their size, owls have large eyes, and their eyeballs are tubular rather than round, a shape which allows for a proportionately bigger retina. This carries a penalty – these tubular eyeballs don't swivel very freely in their bone-plated sockets. An owl can't take a casual sideways glance at something without moving its head. This is where its long and flexible neck comes in. Owls move their heads a lot. They can famously twist their necks in almost a full circle horizontally, and they are also masters of the head tilt, almost to the point that they can look at things upside down.

Another adaptation to seeing in the dark is the tapetum lucidum, a reflective membrane behind the retina which gives owl eyes their eerie shine when caught in a beam of light. This means that light passing through the retina is bounced back, giving the retinal cells a second chance to absorb it. Many other nocturnal animals have a similar structure (and produce a similar eye-shine). The majority of the retinal cells are light-sensitive rods rather than colour-detecting cones. Owls have inferior colour vision to most birds, but when you're operating in the dark, seeing contrast is far more important than seeing colours.

Ears

Birds do not have fleshy external ears. This structure, called a pinna, is the property of mammals only. Some owls look as though they do have visible ears, but these ear tufts are made of feathers alone and have no hearing function (see page 24). However, all birds do have ear openings on the sides of their heads, and similar complex middle and inner ear apparatus to mammals, allowing them to detect sound and also to sense the position of their head in space – the vestibular or balance sense. Additionally, the stiff-feathered ruff or facial disc that most owls possess is there to help channel sound to the ears – the owl can even change its shape as needed by raising or lowering the feathers – so it could be regarded as a kind of 'external ear' after all.

Above: The ear openings are often very large, and are positioned just outside the facial disk on the head-sides.

Below: The shape and sound-gathering power of an owl's facial disk varies according to its state of alertness, like a mammal 'pricking up' its ears.

ANATOMY AND ADAPTATIONS

The long and short of owl ears

Above: Africa's white-faced owls have large ear tufts.

Two of the five British owls have ear tufts. The Short-eared's are just little bumps and often hard to see, but the Long-eared sports a pair of long upright tufts which, in combination with big front-facing eyes, give the bird a definite cat-like look. The same is true of many other owls around the world, including most species of scops-owls, screech-owls and eagle-owls. Ear tufts certainly add to the charm of the owl they adorn, but what is their practical function, given that they are located well away from the actual ear openings and have nothing to do with hearing?

The cat impression sparked the theory that ear tufts are a form of protective mimicry – by resembling a cat, the owl stands a chance of fooling and thus deterring would-be predators. It's an appealing idea, but this kind of mimicry in nature is really only seen where a) there is an obvious 'model'

for the mimic, and b) potential predators are just as likely to approach the model as the mimic, so the mimic benefits by not being attacked by predators that have encountered the model. This isn't the case for cats and owls, as there are tufted owls in many places where no wild cats occur – additionally, many of the eagle-owls are at least as fearsome as any wild cat species that shares their range, while the tiny scops-owls are far *less* fearsome and are readily preyed on by hunters that wouldn't dream of ever tackling any wild cat.

The best explanation for owl ear tufts is that they provide camouflage, by making a jagged outline rather than a smoothly rounded one. At a glance, the owl could be mistaken for a broken branch, especially when it adopts a tall, slim posture, as scops-owls and the Long-eared Owl are particularly inclined to do. Camouflage is needed for even the most ferocious owls – they may not have any predators to speak of, but they may still need to be hard to spot when waiting for prey to wander along. They also like to be left in peace to sleep when at their roosts, and small birds may gang up on and mob even huge owls if they discover them.

Those who work with captive owls have found that the position of the ear tufts can also be reliably indicative of an owl's mood, and suggest they may therefore have a communication function. Alternatively, the link may simply be between the bird's mood and its need to be camouflaged – relaxed and sleepy means the ear-tufts need to be up, for camouflage while roosting.

Above: Lynxes have tufts atop already pointed ears.

Above: Ear tufts create an angular outline.

Owls use their hearing to locate prey. The better and more accurately they can detect and pinpoint the source of sounds, the less they need to rely on their eyes. Using just its keen hearing, an owl can find and catch prey in the dark – but also prey that is hidden in vegetation or even under snow. The Short-eared Owl, our most diurnal owl, still uses its ears to find its prey. It flies low and slow, listening for voles moving along their runs through long grass. It is interesting to note that harriers, which have a similar hunting style, also have an owl-like facial ruff.

Determining the position of an object that's making a sound is accomplished by comparing the timing and relative loudness of the sound as it reaches each ear. This is

something that any animal with two ears can manage easily enough unless the sound is coming from directly below or above – as then the sound will arrive at both ears at the same time, and with the same loudness. Owls – some owls, at any rate – have asymmetric ear openings, one higher than the other, so every sound reaches one ear before the other, regardless of where it's coming from. A handful of species even have the entire inner ears placed asymmetrically, and accordingly unbalanced skull anatomy.

Above: Short-eared Owls hunt by flying low and slow, and listening for prey movements on the ground.

The range of frequencies that an owl can hear is not very different to our own, but owls are particularly sensitive to the sorts of frequencies that correspond to sounds made by their preferred prey. The region of the brain concerned with interpreting sound inputs is also much larger and more complex in owls than in other birds. Experiments with trained, captive Barn Owls have shown they can accurately detect the source of a sound in light conditions so low as to appear completely dark to the human eye.

Touch

Their strange eyeball shape makes owls long-sighted. When dealing with objects at very close range – particularly when manipulating prey or feeding their young – they cannot focus their gaze on what they are doing. They may also need to protect their eyes from a still-living prey item, or a hungry chick, by closing the nictitating membrane - and this doesn't improve their visual acuity. This is when the sense of touch comes in. Owls have specialised fine, hair-like feathers around their bills – these are filoplumes and they function as sensitive 'fingertips', helping the owl get a feel for what's happening right under its nose, so to speak. The rest of the plumage also has scattered filoplumes.

Below: The fine hairs around the bill function like a mammal's whiskers, providing tactile information.

Feathers and flight

Owls spend a great deal of time sitting still, including in daylight hours when colours are much more readily seen than at night. For this reason, they need very good camouflage, and they have some of the most impressively cryptic and detailed plumage patterns of any birds. Most are woodland birds and have plumage dappled with various greys and browns, to help them disappear among branches and against tree trunks. The scops-owls in particular have colours and patterns that are an astoundingly good match for birch tree bark, complete with jagged cross-hatched black streaks that look exactly like cracks and pits in the bark. Among our owls, the Barn Owl is the least obviously camouflaged, but it is also a very active hunter and usually roosts inside buildings or in tree hollows rather than in the open.

Owl plumage is dense and sound-mufflingly soft. The flight feathers in particular are soundproofed in a unique way – their leading edge has a row of comb-like 'teeth' that break up airflow, so the beating wing does not produce the swish of a normal bird's wing but is instead almost completely silent. This means that prey doesn't hear the owl coming, and the owl can focus fully on sounds made by the prey rather than any noise from its own wings.

Above: The plumage is a canvas for camouflage, as in this African Scops-owl.

Below: A close look at an owl's flight feather reveals the 'comb' of sound-muffling barbs.

Curious colours

Above: Barn Owls from different parts of the world vary in colour, but this melanistic bird is much darker than any 'normal' individual.

Right: Abnormal white coloration in an albino Long-eared Owl robs the bird of vital camouflage.

One of the striking things about any long-domesticated bird species – such as pigeons, chickens or ducks – is the variety of plumage colours they exhibit, with white, piebald and unusually dark or even black forms particularly common. Unusual or aberrant colours occur spontaneously in both wild and domestic birds because of genetic mutations, but while these are seized upon and developed through selective breeding in domestic birds, in wild birds odd colours are not usually advantageous at all. Evolution has shaped plumage colours to fit the environment, usually through camouflage, and a white owl in a dark forest is going to have a hard time surviving.

However, many species of owls do have distinct colour morphs, among them our very own Tawny Owl. Most Tawny Owls in Britain have a warm brown colour scheme, but a few are grey. Look at Tawnies in Scandinavia, and you'll find that most are of the grey morph. This fits habitat – in Britain most Tawnies live in deciduous woodland with its richer colours, but in Scandinavia most woodland is coniferous, with trunks, branches and even foliage much greyer in tone.

Above: Tawny Owls of the grey morph are scarce in Britain but more numerous in northern Europe where pine trees are more prevalent.

Birds' wing shapes are adapted to fit certain flight styles. Broad-based and pointed wings, like those on the world's fastest bird, the Peregrine Falcon, generate speed and power. The very long, narrow wings of albatrosses and other seabirds suit sustained gliding flight. Owls have relatively short and broad wings – a shape suited to quick turns and general manoeuvrability, but not so good for power or endurance. Tawny and Little Owls show a more extreme version of this shape, while Long-eared, Short-eared and Barn Owls have rather longer wings. These three are more energy-efficient fliers, tending to hunt on the wing (rather than sit-and-wait hunting, favoured by Tawny and Little Owls) and also covering much more ground in their daily lives – long-distance migrations in the case of the two 'eared' species. A larger wing area relative to body weight means lower 'wing loading' – less energy is needed to power flapping flight. All three of the longer-winged species also make use of the breeze for 'free' lift.

Below: Forest-dwelling owls tend to have short but broad wings – ideal for steering quickly between obstacles.

Behaviour

The day-to-day (and night-to-night) activity of an owl is not so very different to that of other birds. Hunting and feeding takes up some of their time and a lot of their energy. But before we come to that, let's take a look at other important everyday behaviour, including roosting, caring for their plumage, and coping with encounters with other owls and other animals.

Night or day?

We invariably think of owls as nocturnal, but of the British species, three out of five are regularly seen out and about by day. The Short-eared Owl is most likely to be out hunting in the daytime, though only very hungry birds are likely to be up before lunchtime. The Little Owl is also often seen by day, though perhaps doing not very much until the early evening. Males like to 'stand guard' close to their nest holes, though only on days when the weather is good. In fact, male Little Owls

Opposite: A Little Owl attends to its plumage. Preening is an essential daily activity.

Below: Little Owls are often out and about in the daytime, spending much time 'on guard' close to their nests.

Left: Sitting just inside the nest hole is a safe way to enjoy some sunshine.

Above: You are most likely to see Little Owls out in daylight on still, fine days.

Below: The Tawny Owl does nearly all of its hunting by night.

spend so much more time out in the sun than females that it is often possible to tell male from female at a glance when you see a pair – the male is significantly sun-bleached relative to his mate.

Barn Owls are also often seen by day. Both Barn and Short-eared Owls prefer to hunt actively, but the drawback with this is that bad weather can make life very difficult for them, and they may opt to not hunt at all when there is heavy rain or strong winds. As soon as a spell of bad weather passes, they will be eager to resume hunting activity and may then be seen early in the day. Tawny and Long-eared Owls are the closest to strictly nocturnal owls that we have in Britain, though both will hunt by day when necessary. For young Tawny Owls that have left their parents' territory but are yet to find a patch of their own, finding enough to eat can be extremely challenging, and these youngsters are the most likely to be seen out in the daytime.

Roosting and sleep

Sleep is an essential process for all birds, but is also a time when they are at their most vulnerable. Even predators like owls may be targeted by other, bigger predators when asleep, so good concealment at the roosting site is important.

Above: If you are lucky you could spot a Tawny Owl dozing by day in the opening of its nest hole.

Owls may roost inside their nesting cavities, especially just before the breeding season begins, and females incubating eggs or brooding small young have no choice in the matter. Barn Owls and Little Owls often sleep in their nesting holes all year. However, roosting is also often in the open, somewhere close to the regular nest site. Territory-holding Tawny Owls are so loyal to their roosting sites that, once you've found the favoured spot, you can be fairly certain of seeing the owl there every single day. For Long-eared and Short-eared Owls, which tend to move well away from breeding areas in the winter, short-term roosting spots are used, though these will be reused if

Below: Barn Owls like to rest as well as nest inside buildings.

the owl remains in the area and finds the roosting site to be safe and undisturbed. Long-eared Owls in particular will stick to the same area of scrub for their winter roost for weeks or months, though they burrow so deeply into the thorny mass of twigs that they can be very difficult to see even when you know they are there.

From the owl's point of view, of course, ideally you won't find its roosting spot in the first place, as it doesn't want to be found. It will pick places where foliage or tangles of twigs offer concealment. It may roost on a side branch but pressed up against the trunk, so its outline isn't apparent from any angle. Short-eared Owls prefer to roost on the ground, amidst long vegetation that keeps them hidden. Both they and Long-eared Owls

Above: A large winter roost of Long-eared owls in Hungary.

are somewhat gregarious when roosting, gaining safety in numbers. It is unusual for a Long-eared Owl roost in Britain to hold more than half a dozen birds, but in some areas in eastern Europe gatherings of a hundred or more are sometimes found. If you had wondered why on earth there is a collective noun for owls when they are usually so solitary, this is why (and the word is 'parliament' – though this is rather misleading as there is not a great deal of vocal debate, or indeed any other interaction, between the birds in the roost).

Self-care

Above: Frequent careful preening keeps a Barn Owl's wing feathers in flight-worthy condition.

Most birds, including owls, go through a protracted complete moult every year, gradually replacing each of their feathers with new ones. This takes place after breeding. So while an owl replaces its feathers many times in its life, each 'set' needs to stay in good working condition for many months, and so feather care is an important part of an owl's daily routine. Keeping the plumage in good order is vital to health. The fluffy down layer, formed by loose soft barbs at the base of each feather, traps warmth for insulation, while the smooth

Moult

Above: A Cormorant showing wing moult. Nearly all birds replace their flight feathers gradually, so they can still fly throughout the moult.

No matter how diligently a bird looks after its feathers, wear and tear is inevitable. Feathers are quite durable, but they also need to be very lightweight, which affects how tough it's possible for them to be. Colours fade through exposure to sunlight, feathers lose their ability to 'zip' together, and bits of them may break off. Moulting is a physically stressful time for birds – they are not as good at keeping warm or at flying when they're short of a few feathers, and growing new feathers consumes bodily resources. This is why moult nearly always happens when breeding is over but while the weather is still reasonably kind and food quite plentiful. Birds in moult tend to keep

a low profile, and territorial activity is minimal at this time.

Moult of the wing feathers is quite protracted so that flight is not significantly impaired. In most owls, the innermost primary flight feathers go first, one at a time, working toward the outer ones, but in Barn Owls, primary moult starts in the middle and moves in both directions at the same time. In all cases, the owl is never left without most of its flight feathers, though during moult they are a mix of brand new ones and worn old ones.

outer parts of the feathers provide weatherproofing and form the 'canvas' for camouflaged colours and patterns. The long flight and tail feathers provide lift for flight.

Preening involves using the bill to remove dust and dirt from feathers and to get them back into place if they are disarrayed. The main barbs that come out from the central shaft of a feather to form its shape are normally stuck together through a Velcro-like arrangement of tiny hooks, but if the barbs become separated, preening helps to 're-zip' them. If a feather has become detached completely but remains caught on surrounding feathers, preening allows the owl to dislodge it.

Most feathers on an owl's body are accessible to its bill – having a very flexible neck certainly helps here. The only parts that are not are the feathers on the face and head. Owls can deal with these by scratching with their talons, though this is rather less precise than preening with the bill. Owls can, of course, preen each other's faces with their

Above: The famously flexible owl neck is helpful in a thorough preening session.

Below: Owls bring a foot directly to the head to scratch, unlike some other birds which lift the foot over a lowered wing.

Above: The best way to look after one's head plumage is to ask one's partner to preen it.

Below: A Great Grey Owl takes a bath, using its wings to throw water across its back and head.

bills. Much better than a scratch is a lengthy session of mutual preening, and this is one of the many reasons why owls tend to form lasting pair bonds with a life partner. In some species, including the Little Owl and its much more sociable American cousin the Burrowing Owl, fledgling owlets are quite friendly with each other and will also attend to each other's facial preening needs.

Like most birds, owls enjoy a bath in water from time to time. They stand in shallow water and bob and dip, using their wings to shake the water over their plumage. They will also dust bathe, finding an area where there is loose dust or sand on the ground, and making similar movements to shake it through their feathers. Both kinds of bathing help dislodge anything stuck in the plumage, and are followed by a lengthy preening session. While bathing they are vulnerable to predators, so baths need to be short and sweet, and ideally done somewhere where all-round visibility is good.

Dust and oil

Some birds possess powder down – a kind of self-grown cleaning product. Specialised powder-down feathers crumble at the touch, yielding a powder, which the bird spreads through its feathers to absorb dirt and grease. Herons, which deal with messy aquatic prey on a daily basis, are the best-known bearers of powder down. It is not certain whether owls possess it or not – some authors assert that they do but the evidence is not conclusive.

Owls do produce preen oil, a waxy substance that comes from the uropygial gland at the base of the tail. This oil is collected on the bill and then spread through the plumage. Its main purpose has long been supposed to be waterproofing, but it may simply help improve feather integrity by improving flexibility and so reducing the chances of breakage. There is also growing evidence that its chemical composition discourages parasites.

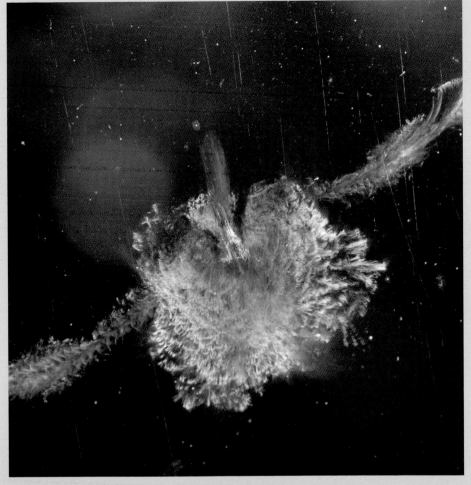

Above: A 'ghost' in powder down, left on a window after an avian collision.

Social life

Above: Little Owl chicks associate closely with their siblings for some weeks after fledging.

In general, owls are like most other predators in that they lead rather solitary lives. Pairs may live in the same territory long term but they don't hunt together. They may roost quite close together but don't have much to do with each other when not breeding. Long-eared and Short-eared Owls, which move away from their breeding grounds in winter, do form communal roosts at this time, as previously mentioned. They benefit from shared vigilance against danger and they may also observe flight paths of other owls as they arrive or depart, and investigate those directions as possible hunting grounds, but there is no real cooperation going on. Sometimes multiple Short-eared Owls may be seen hunting over the same field, and being broadly tolerant of each other, though they will have aerial tussles when one encroaches too far on another's space.

All of the British owl species may at times hunt other birds, and all may in turn be mobbed by other birds. Mobbing is systematic harassment, with the goal of

forcing the predator to move away. Small birds will gang up to dive-bomb a roosting Tawny Owl, giving loud alarm calls which draw even more mobbers to the scene. Mobbing usually happens at a time when the predator is trying to sleep or rest, so is unlikely to turn on its tormentors. This is a key reason why owls' roosting spots need to be well hidden. The most frequent victims of mobbing are young owls dispersing from their parents' territories – they are searching for a home of their own but the local small birdlife will be keen to discourage them wherever they go.

Larger birds may harass owls too. Corvids (members of the crow family) are too big to be easy prey for any of our owl species, but their fledglings could be potential prey, so they show a general intolerance towards anything with a hooked bill. Any uproar among corvids should be investigated as it often signals the presence of a bird of prey, though that bird is much more likely to be a raptor than an owl.

Above: A Hooded Crow harasses a Eurasian Eagle-owl, in the hope it will find another place to roost (and, later, hunt).

Territory, Competition and Migration

Life without a permanent territory is difficult, if not impossible, for a Tawny Owl. For a Short-eared Owl, territory is a temporary and transient thing. Most of the owls in the world follow the sedentary Tawny's mode of life, rather than the nomadic Short-eared Owl's, but the Short-eared is arguably the most successful of all owl species – certainly, it has an exceptionally wide global distribution. However, all owls agree that a territory of some kind is vital during the breeding season, as a brood of chicks need a nest and a reliable supply of food nearby for their parents to bring to them.

A home for life

Tawny Owls are the most territorial of all our owls, and the least keen to travel, because life on the move is very difficult for them. Young owls dispersing from 'home' at the end of summer will always try to find a territory very close to where they were born – the longer they are territory-less, the slimmer are their chances of survival, especially as winter kicks in. Ringing studies organised by the British Trust for Ornithology (BTO) are our main source of information on birds' movements. Trained ringers trap wild birds (or take chicks from the nest) and fit them with unique numbered rings. This means that ringed birds can be individually identified if they are caught again by ringers, or found dead. The BTO's data on ringing recoveries for Tawny Owls show that they move an average of just 4km (2½ miles) over their lifetime – typically in one trip from their birthplace to wherever they manage to establish a territory – or die in the attempt.

Opposite: Little Owls are home-bodies, and will not leave their territories if they can help it.

Below: Facing its first winter, a young Tawny Owl desperately needs a safe hunting territory if it is to survive.

Tawny travellers

Above: Fitting a leg ring to an owlet.

A handful of ringed British Tawnies have been found to have made long-distance movements in their lifetimes. The one that went furthest was hatched in the Scottish Highlands, and killed by a car six months later in Wales, having travelled 688km (428 miles). This bird stands alone as an adventurer – its closest rivals covered less than 300km (186 miles). Another of the adventurers was luckier – ringed as a chick in Nottingham in 2007, it was retrapped in 2014 in Northumberland, alive and well and on its breeding territory.

The most startling recovery was of a ring placed on a Tawny Owl in Britain, and then found in Iceland. Not only would the journey of more than 800km (500 miles) be unprecedented, but (as we know by their failure to colonise Ireland), Tawnies are not keen on even short sea crossings. Indeed, no other British Tawny had ever been

known to leave Britain before. The fact that the ring was found, but not the bird, was a clue to what had actually happened, and it transpired that the poor owl concerned had actually died in the nest so had travelled no distance at all. The ring had been taken off the dead owlet's leg by the ringer, to save erroneous future reporting, but then the ringer had lost the ring on a birdwatching trip in Iceland.

Another ringed Tawny met a notably unfortunate end – its notes state that it was recovered in August 1999 after being 'possibly hit by a tennis ball'. This bird did enjoy a long life, having been ringed as a chick in 1983. Perhaps as an old bird it was ousted from its territory by an incomer, and was wandering in search of a new patch – flying tennis balls are not a usual hazard in a well-appointed Tawny Owl territory.

The Tawny Owl defends a territory of anywhere from 20ha (50 acres) to more than 400ha (990 acres), with smaller territories in denser woodland. Paired birds' territories overlap extensively but not completely, especially in winter. Its extreme reliance on its territory is related to its hunting style. Experience is everything – this owl usually hunts by waiting on a favourite perch and pouncing on prey that wanders by. Not every perch is suitable for this, and from the same perch, not every flight path will be suitable. The Tawny Owl needs to learn where the best places are and how best to use each one. It also works out where to find particular prey items – damp spots where earthworms come to the surface, shallow waters where fish and frogs can be caught, hedgerows where roosting songbirds may be snatched at dusk. Over time it builds an intimate familiarity with every detail of its territory, and in effect, the longer it lives in its one particular patch, the

Above: A Tawny Owl knows its territory inside out, down to the best spots to find the biggest earthworms.

more skilled it becomes at living there. The territory must also have a good place to nest, and for these rather large owls, suitable cavities are quite rare.

No wonder, then, that Tawnies guard their territories with considerable ferocity. A young Tawny has very little chance of evicting another healthy adult Tawny from a territory. Other birds of prey are also given short shrift. Tawnies will drive away and can kill Sparrowhawks, the day-flying raptor most likely to occur in their territories,

Above: A good nest site is priceless for Tawny Owls and will be used year after year.

and they will do the same with Long-eared Owls – competition from smaller relatives is very unwelcome. (On the Continent, the Tawnies may themselves be targeted by larger owls, such as the Eurasian Eagle-owl and Ural Owl, but in Britain they are very much 'top owl'. However, the Sparrowhawk's big brother, the Goshawk, can kill Tawnies.)

How does a young Tawny Owl go about finding itself a breeding territory, when the chances are that all the good spots are already held? Perhaps its best bet is to eke out a living on the fringes of other pairs' territories, and bide its time in the hope that an incumbent bird of its own sex dies, or becomes weakened through age or accident and therefore potentially beatable in a territorial battle. Finding a place that will do as a hunting territory is not so difficult, though, as that only needs to provide food and shelter, but a breeding territory must have a nest site – the rarest of the resources a Tawny needs.

If the owl is very lucky it may have a nest site handed to it on a plate – for example, if a tree is damaged in a storm, leaving an accessible hollow, or the landowner erects a nest box. Young owls are able to breed at one year old, but competition for territory may mean they do not get the chance for three or four years, because of the lack of a good enough territory. Once a male has established a territory, he will begin to advertise (through song) for a female, and a young female in her first territory will listen for available males nearby.

Wandering ways

Truly nomadic behaviour is rare in birds. Many species regularly migrate, but usually winter in the same areas and return to the general area where they were born when it's time to breed. A nomadic species, though, won't necessarily do this, but could breed and winter in different locations from year to year.

Ringing studies on Short-eared Owls show that long-distance movements and sea crossings are par for the course. Nine Finnish-ringed Short-eared Owls have been recovered in Britain, while six British-ringed birds were found in Spain. The longest journeys include several

Above: Migration may take a Short-eared Owl to unexpected places; here the German seaside city of Cuxhaven.

Below: Many owls are reluctant to make sea crossings, but the ocean is no barrier to the Short-eared Owl with its wandering ways.

Above: The pursuit of voles drives the long-distance migrations and wanderings of Short-eared Owls.

of more than 2,000km (1,240 miles), involving birds travelling between Britain and Iceland, Morocco, Russia and Malta.

Short-eared Owls take a flexible approach to life because of their reliance on one particular kind of prey – voles, especially grassland voles. In Britain and western Europe, that means Short-tailed Voles. These little mammals can make up more than 90 per cent of Short-eared Owls' diets. Like many other small rodents, they tend to go through population booms and slumps, and this directly impacts their predators. Short-eared Owls move to where the voles are, so breeding areas and particularly wintering areas vary year on year. In the winter of 2015–2016, thousands of 'extra' Short-eared Owls were in Britain – birds from the Continent, forced to wander far and wide in search of voles. The rodents tend to recover from a slump very quickly, having a prodigious rate of breeding. Short-eared Owls can take full advantage of a vole glut by producing bumper clutches of eight or more eggs in a good season, rather than the more typical five – they may also produce two broods in a season, while the Tawny only ever has one. The Short-eared's fast pace of life may take its toll on its lifespan, though – the oldest ringed British Short-eared Owl was six years and eight months old, but the oldest British Tawny was almost 22 years old.

The middle ground

The three remaining British owls fall in between the ultra-sedentary Tawny and the freely roaming Short-eared in their attachment to territory and propensity to wander. Little Owls are highly sedentary, moving on average 15km (9⅓ miles) in their lifetimes, and like Tawnies they avoid sea crossings – hence the failure of the Little Owl to colonise Britain from Europe under its own steam. No British-ringed Little Owl has ever been recovered abroad, and their longest recorded movements are all under 200km (124 miles).

This owl stays with its mate year round, both birds defending their shared territory, which usually covers 20–40ha (50–100 acres). The pair uses the same nest site year on year if they can. Being smaller, they often have more choice of nest sites than Tawnies do – they can make use of old Green Woodpecker nest holes and are also happy to use cavities in buildings – and they

Below: Paired Little Owls can be seen at their favourite spots in their territories all year round.

Above: Though most owls prefer to nest in tree holes or rock crevices, many will also use buildings.

are also less numerous than Tawnies, and shorter lived. Therefore, young Little Owls probably have a slightly easier time finding a suitable territory than do young Tawnies. However, their territories have more flexible boundaries. The area they defend expands in winter, and contracts in the breeding season when caring for chicks is a higher priority than guarding the borders. In times of severe winter weather, though, Little Owls need to devote all their time to hunting and will tolerate intruders much more, so effectively their territory shrinks.

When it comes to the nest site, Little Owls may have to fend off other hole-nesting bird species, ranging from Stock Doves and Jackdaws to Kestrels, Barn Owls and (in some areas) the introduced and surprisingly feisty Ring-necked Parakeet. Cameras fitted in nest boxes have caught some ferocious battles between various combinations of these species. Usually, the bigger and more predatory species wins, but in one filmed encounter a Little Owl that was already incubating her clutch successfully fought off an intruding Barn Owl.

Long-eared Owls are much more similar to Short-eared Owls, being willing and able to undertake long journeys when necessary, and usually moving away from breeding areas in winter. However, they are not truly nomadic and return to their regular breeding areas in spring, occupying the same territory for breeding each year if they can. Although they do hunt voles, they are not as reliant as Short-eared Owls on this particular prey type, and hunt plenty of birds as well as other small mammals. This flexibility in diet means they can afford to be a little less flexible in how and where they live, and enjoy some of the benefits (as shown by the long-lived Tawny Owl) of regular territory occupancy. The longest travels of ringed Long-eared Owls in Europe mainly involve birds crossing the North Sea from north-west Europe, especially Scandinavia. Birdwatchers 'working' the east coast of Britain in autumn regularly observe Long-eared Owls arriving on the shore, and sitting incongruously on the beach for a rest before continuing inland in search of more suitable wintering habitat.

Above: Many Long-eared Owls in Britain during winter have come here from colder north-east Europe.

Above: Many Barn Owls die in severe winters, as they live in open where hunting – and sheltering – is difficult in bad weather.

Below: The Long-eared Owl prefers to eat small rodents but will switch to other prey when conditions make rodent-hunting difficult.

Barn Owls are mainly quite sedentary. A lot of nature reserves nowadays have one or more Barn Owl nest boxes, and if you train your binoculars on the entrance hole of an occupied box, you will often be able to see the sleepy face of one of the pair inside, regardless of what time of year it is. If a Barn Owl pair has a productive territory with a reliable nest site, both birds will stay put and defend their patch all year round, just like Tawny Owls.

However, conditions in the open countryside that Barn Owls like are much more variable than in the interior of a woodland. Tree cover offers a lot of protection from the vagaries of the elements – conditions like flooding, snowfall and gales have a lot less impact on a Tawny Owl's hunting options in the shelter of the wood than for Barn Owls hunting over exposed open countryside. These owls can struggle to hunt in bad weather of all kinds, and they are also quite reliant on voles as a food source (not quite to the same extent as Short-eared Owls, but enough that vole population crashes can have a serious impact). They can compensate for hunting troubles to some extent by going out in the daytime if they fail at night, but sometimes the only option is to give up on the territory and move further afield. Movements of several hundred kilometres (a few hundred miles) are not unusual, and there are many recoveries of British-ringed Barn Owls

from mainland north-west Europe and vice versa. The ringing recovery data also includes one extreme outlier – a bird ringed as a chick in Oxfordshire in summer 2005, and then found dead the following spring in Afghanistan, 5,834km (3,625 miles) away. Why the bird undertook this epic journey will always be a mystery, but the fact that it did shows that Barn Owls are capable of travels that are beyond many a truly migratory species.

Above: Derelict buildings in the countryside offer vital shelter as well as nest sites for Barn Owls.

Shouting from the rooftops

How do owls defend their territory? In much the same way as other birds – with song. The males of all of our owls have territorial calls which have the same function as songbirds' songs, though they may be less melodic. The song serves to let other owls know that a particular patch is occupied, and if the singer is a singleton it also works as an advertisement to potential mates. A female may move wholesale into the male's territory, or if they are close neighbours their territories may be combined and the boundaries redefined.

The Tawny Owl, as our most territorial species, is also the biggest singer, and the male's long, quavering, fluting hoot is a familiar sound in many areas on clear nights, especially in late winter and spring. Paired birds will 'duet', both birds giving

Left: The quizzical squeaky yelp of a Little Owl is quite different to our other species' calls.

Below: Tawny Owls have a wide repertoire of calls but their fluting hoot is the best known.

Above: In moments of high emotion, owls' calls become higher pitched and more frequent.

the same fluty hoot, the female's a little higher pitched. This duetting helps with pair-bonding but also shows a united front against all intruders – the female is as keen to keep away other females as the male is to deter other males. The familiar sharp 'ke-vick' call (the 'to-whit' part of the 'to-whit to-whoo' rendition of Tawny calls) is a contact call used by either sex, most often by a female in response to her mate's hoot.

Long-eared and Short-eared Owls are also hooters, the Short-eared giving a series of short, pumping hoots and the Long-eared a longer note, but both lacking the ethereal flutiness of Tawny hoots. Both species also use display flights with wing-clapping as part of territorial defence and mate advertisement. The Little Owl's hoot is so high-pitched as to be almost a yelp and becomes very squeaky indeed when the singer is highly agitated. It has a ringing quality and often an interrogative, upward inflection. The Barn Owl advertises its territory with its famous hissing or grating screech, given repeatedly.

Above: Barn Owls produce a variety of snorts, wheezes, hisses and shrieks.

Diet and Hunting

Owls of all species are predators and make their living by catching, killing and consuming other animals. Between them, they eat a vast array of different kinds of animals. Some of the smaller owls (and even a few of the bigger ones) tackle nothing more challenging than beetles, moths and worms. In fact, more than half of the world's owls are primarily insect eaters, but most species (including all five of the British owls) take at least some vertebrate prey, ranging from fish, frogs and lizards to small and medium-sized birds, and mammals as big as deer fawns and Foxes. No animal wants to be eaten, and those at the larger-brained end of the scale, in particular, are quite good at avoiding it. Owls need to deploy guile, stealth and sometimes dirty tactics to overcome their prey's defences.

Diet

We think of owls as eaters of mice, and certainly, all of our owl species will cheerfully dine on mice if they can catch them. But only the Short-eared Owl and to a lesser extent the Barn Owl are particularly fixated on eating rodents, and their rodent of choice is the Short-tailed Vole, also known as the Field Vole. This little 30g (1oz) parcel of protein is the most abundant mammal in Britain, with a population estimated at 75 million (though subject to considerable fluctuation, tending to rise and fall on a four-year cycle). It is a grassland vole, unlike Britain's other common small vole, the Bank Vole, which lives in woods and hedgerows. A female Short-tailed Vole can produce 25–30 young in a year – none of these are likely to live for longer than a year themselves, but this impressive fecundity is what allows the vole population to recover quickly after numbers crash. Why such crashes occur is not certain, but cyclical population booms and busts are seen in a number of other northern rodent species, and they cause similar population effects in their key predators.

Opposite: Barn Owls have long been valued by farmers as controllers of mice and rats around grain stores.

Below: The most versatile of our owls in dietary terms, the Tawny is best known as a hunter of rodents.

Above: From an early age a Tawny owlet can swallow a vole in one go.

Other small British rodents are also predated by owls, as are shrews. Every small mammal species found in Britain, including rarities like the Hazel Dormouse and localised species like Water Shrews, has been recorded as prey for the Tawny Owl, and these powerful hunters also take the occasional squirrel, infant Rabbit and

Right: Hunting by sound allows Short-eared Owls to locate voles even under snow cover.

Weasel. Urban Tawnies take large numbers of Brown Rats and House Mice. Barn Owls prey on House Mice and juvenile Brown Rats, too, where these are abundant around farms (rather a rare scenario these days), as well as sometimes on baby Rabbits. Owls may also prey on bats. Little Owls are mainly insect eaters but even they are quite capable of dispatching small mammals if they manage to catch them.

All of our owls will also hunt birds from time to time. Even the Short-eared Owl, with its vole-heavy diet, will catch a few birds, particularly in the breeding season when fledgling Skylarks and Meadow Pipits are not yet flying and are running around in the grass. Barn, Tawny and Long-eared Owls are all quite skilled at catching songbirds as they go to roost, and may fly close along hedgerows in the hope of flushing them out. Tawnies will occasionally rob other birds' nests, attacking both chicks and incubating adults. Long-eared Owls probably take · the most birds of all our owls, but there is considerable variation from season to season and between different habitat types.

Above: Little Owls take many insects, though learn to avoid bad-tasting ones like this burnet moth with its bright warning coloration.

Foreign food

Left: A Galapagos Short-eared Owl with its storm-petrel prey.

Below: In captivity, nearly all owls will thrive on a diet of day-old chicks – an abundant byproduct of poultry farming.

Three of our five owls occur over a broad swathe of planet Earth, and different populations eat different types of prey. The Short-tailed Vole is only found in Eurasia, so Short-eared Owls elsewhere in the world eat different kinds of small rodents. Mainly these are voles that are of similar size and have a similar ecology to the Short-tailed Vole, but a study in Texas found that the favourite prey there was the much larger Hispid Cotton Rat. Wherever they live, Short-eared Owls seem to specialise in the most abundant local grassland rodent – and everywhere has one of those. Or nearly everywhere … there is an exception, of course. The Galapagos Islands have very few land mammals, and the Short-eared Owls there have switched to a diet of lizards and birds – particularly storm-petrels. These small seabirds return to their burrows at dusk to avoid predators, but the ones on the Galapagos Islands find a welcoming committee of owls waiting to try to grab them. (The Short-eared Owls of the Galapagos have also discarded their species' nomadic ways, and in fact are now so different to other Short-eared Owls that they are sometimes considered a separate species.)

Long-eared Owl diet is usually mainly composed of small mammals but can be quite variable, even between populations that are close neighbours. Two studies in different parts of Israel found a striking difference – in one the diet was 92 per cent birds (by weight) but in the other mammals dominated at 71 per cent. Our two 'eared owls' have taken the world by storm as a result of their feeding habits – one wins by specialism, the other by generalism.

Barn Owls are mainly mammal hunters as well – in Britain and other temperate areas voles and other small mammals usually make up at least 90 per cent of the diet. However, Barn Owls elsewhere in the world mix things up a bit – those living in hotter and more forested areas take insects and lizards in large numbers. In a few areas where mammals are scarce, such as the Cape Verde Islands, birds are the main prey. An ornithologist visiting a tiny rocky islet off California in 1928 found a Barn Owl nest there, with chicks that were being raised (going by prey remains) on a diet consisting exclusively of Leach's Storm-petrels.

Our two most sedentary owls are also our most catholic in terms of diet. This is a predictable trade-off – either travel to find the food you like, or stay where you are but accept whatever food is going. Tawnies eat large numbers of earthworms, and also readily take amphibians, fish and will even eat carrion. Little Owls eat insects and other invertebrates of all kinds, and anything larger that they can catch – in fact, vertebrate prey can make up half or more of the diet by weight. Although small, the Little Owl is a tough little predator and can catch birds close to its own weight and even kill a young Rabbit.

An interesting phenomenon that can be observed in owls and day-flying birds of prey alike is that of intraguild predation. A 'guild' is a group of species that coexist and compete for the same resources – and intraguild predation is when these species eat each other. There is broad overlap in the diets of all of our owls – so if a Tawny Owl kills a Long-eared Owl, it not only gets a meal, but also removes a potential competitor from its habitat. Hunting other predators is more difficult and dangerous than sticking to conventional prey, as predators are harder to catch and more likely to be able to defend themselves, but the bonus of eliminating competition outweighs the effort and risk. Intraguild predation is an important factor in determining which predators are found where, and their relative abundance.

Below: Mammalian predators tend to avoid eating shrews, but owls have no such qualms.

Studying owl diet

It is not surprising that there is a large body of research on owl dietary composition, as owls are an absolute gift to anyone interested in studying predator diet. No need to observe them hunting or feeding, or even to analyse their droppings, because owls cough up pellets that contain large and easily identifiable bits of whatever they have eaten. Moreover, they cast their pellets where they roost or nest, so once you have found these places you can be reasonably sure that you will find all the pellets.

Because owls swallow their prey whole, the pellets usually contain intact skulls and other bones of vertebrates, bent but not broken feathers of bird prey, and intact wing cases and other hard parts of invertebrates. Pellets can be teased apart easily to access their contents after soaking in hot water with a little disinfectant (in fact, this is sometimes recommended as a 'nature detective' activity for children, though it must be done under adult supervision and with rigorous hygiene). Skulls are the 'best' items to find as they are easier to identify to species

Above: Owl pellets look like clumps of matted hair, but hidden within are many small bones.

Below: Biologists can use owl pellet analysis to work out which small mammals occur in which areas.

than the more generic bones of other body parts. Once you have sorted out which species' remains are in the pellets, you can then work out what proportion of diet (by number, and then by weight) is made up of which species.

Hunting methods – action or patience?

As mentioned already, there are two ways to obtain your dinner – you go out and search for it, or you wait for it to come to you. The second of these is not an option at all for most animals as their food tends to stay in one place, but for owls, which hunt active and mobile prey, it is. It might take a bit longer than actively searching for prey, but this is offset by its minimal energy cost. In fact, hunting by waiting is very popular among owls and is one reason why it is difficult for people to see them – if you don't know the owl is there, neither does its prey.

Below: Tawny Owls are sit-and-wait hunters, sitting on favourite perches and constantly scanning for signs of prey.

Searching on the wing

Although owls are mainly sit-and-wait hunters, three of the five British owls regularly hunt on the wing, covering considerable ground as they fly low and slow over suitable terrain, ears and eyes on full alert for signs of little life forms moving below. The Barn Owl, Short-eared Owl and Long-eared Owl all have long and large wings relative to their body weight. This very low 'wing loading' maximises

Above: Constantly on the move, a hunting Short-eared Owl patrols a field in a 'quartering' search flight.

the energy efficiency of these species' slow, flapping flight. The famously silent flight ensures prey remains oblivious to the owl, and that the owl itself is not distracted by the sounds of its own beating wings. When the owl hears or sees prey, it often hovers on the spot momentarily, head down to line up its strike. Then it drops like a stone, feet first, hopefully landing with talons on target.

This method of hunting is used by some of the diurnal raptors as well, most notably the harriers. At the right time and place in Britain, you could observe Short-eared and Barn Owls and Hen Harriers all 'working'

the same area of rough grassland. Weather conditions affect the ease and likely success of flight hunting for all of these birds, and dry weather with a bit of a breeze is ideal. The birds can then fly into the headwind and use its uplift to save a little energy. If you are watching an actively hunting owl, try to position yourself with the wind behind you, and the bird may come towards you. Very still air makes active hunting harder work, and the same is true of very windy and rainy days. When the weather is unsuitable, the owl may resort to sit-and-wait hunting from a perch instead.

Above: Barn Owls often hover before pouncing, listening carefully to pin down the exact position of their prey.

Above: Little Owls are more confident on their feet than our other species.

Run for it

With their zygodactyl feet (two toes pointing forwards, two backwards) and big curved talons, owls are much better at gripping a branch (or hapless victim) than they are at standing and moving on flat ground. The Little Owl is something of an exception, being a competent runner, and it often chases prey such as large insects on foot. Its close American relative, the Burrowing Owl, is the most terrestrial of all owls and spends much of its time on the run, whether dashing along in pursuit of prey or bolting back to its burrow to escape from danger.

Right: Hunting on the ground, a Little Owl runs and jumps after its prey.

Wait for it

In the 'cluttered' environment of a woodland, hunting by flying about has obvious drawbacks. Manoeuvring around a constant series of obstacles is much harder work than flying unimpeded in open countryside with the help of a breeze for uplift. Most owls do live in such cluttered environments, however. Some of them move out of the woods to hunt actively in adjoining open areas but others stay in the woods and use sit-and-wait or 'still' hunting instead. The Tawny Owl is a classic example of this kind of hunter.

Steal it

Kleptoparasitism is the phenomenon of one animal stealing food directly from another. The masters of this are the skuas and the frigatebirds, seabirds that are expert chasers and harassers and obtain a large proportion of their food by theft. The behaviour is also seen in landbirds, including owls. Short-eared Owls working over a field in a group might try to swipe voles from each other's talons. However, owls are more likely to be victims of this thievery than perpetrators. Kestrels are particularly skilled food thieves and will target Barn and Short-eared Owls.

Above: Perfect hunting perches for a Tawny Owl offer a good view and are not too high up.

Left: Owls are more often victims than perpetrators of prey theft. Here a Kestrel tries to snatch a vole from a Short-eared Owl's talons.

Pairing and Breeding

Raising a family is a two-owl job, from start to finish. Owls, therefore, form lasting pair bonds, and the sexes have equal (though different) roles at all stages of the process. Broadly speaking, the male is the provider and the bigger, stronger female is the protector. Courtship involves the birds testing out each other's relevant skills, and building up a relationship based on cooperation. When eggs and chicks come along, those skills and that ability to work together are put into practice.

Perfect partner

In nature, it is usually the female who chooses a mate from the males available. This is because biologically she can produce fewer offspring – her egg cells are big and more energetically costly for her body to produce, and relatively few can be made in her lifetime. Lower reproductive potential makes it more important to choose the 'right' partner with which to reproduce. Males, who can produce lots of small, 'cheap' sperm, have the potential to have many more young so 'should' be (and often are) more concerned with mating with as many females as possible to maximise their reproductive output.

Where this falls down is parental care. For owls and many other birds both parents need to be around until the chicks are independent. One needs to stay at the nest to incubate the eggs, brood the young while they are still small and helpless, and generally guard the nest, while the other goes out and fetches food for the whole family. (The exceptions are birds in which the young are fully 'precocial' – able to run about and feed themselves within hours of hatching. Some of these species, which include some gamebirds and wildfowl, can make single parenthood work for them.) So even though a male owl could theoretically sire numerous owlets with lots of different females, he can usually only commit himself to one partner and one nest per season.

Above: Female Barn Owls are usually a little darker and more spotted on the breast than males.

Opposite: A Little Owl preens its mate's head feathers in a touching display of closeness and devotion.

Above: Mating Little Owls, as with other birds, need to employ both strength and balance.

Choice does still tend to come down to the female, though, as she is the one who responds to a male's territorial advertising calls at the start of the breeding season. A male owl's preference is for a bigger and stronger female, but given how hard-won territory can be, when a male has secured a good one he will not leave it to go in search of females. So he will accept whichever female comes to him, and he does risk not attracting any at all. But this is a much smaller risk to his lifetime reproductive potential than leaving his territory would be, as if he did so he would probably lose it to another male, and have nothing with which to attract a mate. This will also harm his own chances of survival – especially in the case of the sedentary and highly territory-dependent species.

When a female is investigating local unattached males' territories she will choose the 'best' male she can find, with the highest quality territory. If two females are interested in the same male and his territory, he is happy to leave them to sort it out between themselves, as the victor will have demonstrated her tenacity and ability to fight and win.

Opposite: When the male is balanced properly, the female raises her tail to allow him to bring his cloaca in contact with hers.

A song and dance

Above: Territorial song is vital to attract a mate and to warn off potential rivals.

The importance of vocal communication to owls can't be overstated. In most species, hearing is the dominant sense, and in a dense woodland there is no better way to make your presence known to other species, particularly at night, when most other birds are quiet. Nearly all owls have distinctive and species-specific territorial songs, to the extent that some very similar species are separated on the basis of their different songs, when the birds themselves are nearly identical in appearance. Once a male's territorial song has attracted a mate, the pair will sing together, and this is an important part of the courtship and bonding ritual.

Short-eared Owls, which are active in the day and are also birds of open rather than forest habitats, use visual display as part of their courtship. The male's territorial display flight is spectacular to see. He flies high, much higher than in typical hunting/searching flight, with very deep wingbeats. He then suddenly tumbles steeply earthwards, with loud wing-clapping, pulling up just a metre or two (a few feet) above the ground. All of this is accompanied by calling. Like other owls' songs, the display serves to deter other males and to attract females. Long-eared Owls also perform a courtship flight with wing-clapping – the sound rather than the visual show is more important in this case, as it takes place at night.

Bonding

Direct interactions help a new pair get to know each other, and a reacquainting pair to remember each other. Even owls that form long-term pair bonds drift apart somewhat after breeding, and rituals like mutual preening, duetting and courtship feeding (where the male brings food for the female) help to unite them and build their shared focus on the tasks ahead, as of course does mating.

Owls that are sedentary usually keep the same partner year after year. Proven breeding success heightens their motivation to stick together. The species that roam around in winter – Long-eared and Short-eared – will not necessarily have the same mate in successive years, but if both birds from a successful breeding pair survive and return to the same breeding area, they will seek out the same territory as before, and so may well end up together again by default.

Below: A male Short-eared Owl approaches with care as he delivers a meal to his hungry mate.

Nesting without a nest

Some birds are master crafters, using raw materials from the world around them to build intricate, robust and beautiful structures in which to lay their eggs and tend their young. Owls are not in this category – no owl builds anything that could be regarded as a proper nest. So they save themselves a good deal of hard work, but they still need a safe place for eggs and chicks and really suitable places can be hard to find.

Above: Grey Herons build large solid nests from sticks, and such nests may be reused by some owl species.

Holed up

A cavity in a tree trunk is a fine nest site for an owl. It offers shelter from the elements and is quite difficult for predators to access. There are two kinds of tree holes – those that form naturally, for example, when a branch breaks from the trunk and decay hollows out the resultant wound, and those that are made on purpose by other birds. The latter make particularly good nest sites, as they were created for that very purpose. Birds like woodpeckers hack out cavities in which to nest and often abandon them after one breeding season, leaving the way open for a pair of owls to move in. (And of course, if the original owners are still in residence they may be encouraged to vacate by the owls!)

In Britain, woodpecker holes are not much utilised by owls – our woodpecker species are all a bit too small and our owls too big. A Green Woodpecker nest hole may be suitable for a pair of Little Owls, but the old nest holes of Great Spotted Woodpeckers are likely to be too tight (and those of Lesser Spotted Woodpeckers certainly are). The most woodland-dependent of our owls, the Tawny Owl, will usually nest in a natural tree hollow, and the relative scarcity of such hollows is one reason why Tawnies are so steadfastly loyal to their territories (another is the importance of great familiarity with their surroundings from a hunting perspective – see page 43).

Opposite: The perfect tree house for a Tawny – the hole left where a branch has fallen and the exposed wood has rotted away.

Back off

Above: Large owls like the Ural Owl can be violently protective of their nest and young.

If you discover an owl's nest site, you may naturally want to watch proceedings through the season, but it's important you do so from a respectful distance, to avoid disturbing them. The Barn Owl is also on Schedule 1 of birds that have enhanced legal protection, making it an offence to disturb it at or near the nest. It's also important for your own safety to keep your distance because owls can be fiercely protective of their nests. Tawny Owls, in particular, may launch aerial attacks on humans who stray too close. The bird photographer Eric

Hosking lost an eye as a result of an attack by an irate Tawny defending its nest. The Ural Owl, a supersized version of the Tawny that is found over much of Eurasia, is notorious for attacking anyone near its nest and is known in Sweden as *slaguggla* ('strike-owl'). Ringers visiting its nests wear head and eye protection, and take a friend along to fend off the adults if necessary with a long branch.

Crevices in cliff or rock faces may also be suitable nest sites, and Barn and Little Owls, in particular, will use holes in buildings. Barn, Tawny and Little Owls are also very happy to use purpose-built nest boxes. If properly built and carefully sited, a nest box can be better than any natural nest site and their widespread provision is one of the reasons that our Barn Owl

population has increased over the last couple of decades. Elsewhere in the world, nest-box provision is helping to save the endangered Blakiston's Fish-owl, an East Asian owl so huge that it struggles to find a natural tree hollow large enough for it, and in Scandinavia providing nest boxes for Ural Owls has helped the species to thrive alongside forestry operations.

Above: A gap in the brickwork of an old rural building is a very suitable nest site for Little Owls.

Born in a barn

Birdwatchers are often asked where Barn Owls lived before there were barns. The answer is that they would have nested in other hollows, most likely in tree trunks or rock faces, but accessible farm buildings with shelves or ledges inside offer ideal accommodation for this open-country owl. The recent trend to 'do up' such buildings and convert them for residential use combined with the clearing of dead trees has reduced the number of good nest sites accessible to Barn

Above: Before barns, Barn Owls would have nested in tree holes and caves.

Owls, though these have perhaps had less impact on their numbers than other factors, such as increased pesticide use. Rural landowners can easily address any housing crisis among their local Barn Owl population by placing nest boxes on or inside undisturbed farm buildings – the reward could be weeks of enjoyment watching a pair of these lovely owls successfully rearing a family.

Second-time buyers

Tree holes are scarce and getting scarcer. Our idea of sensible woodland management is, all too often, based around chopping down the mature and half-dead trees that are most likely to offer owl-friendly hollows. For Long-eared Owls, which in Great Britain tend to be pushed out of the 'best' mature woodland by the presence of the bigger Tawny Owl and, instead, live on woodland edges and in young plantations, it's hardly worth even trying to find a tree hole. Instead, Long-eared Owls frequently use old nests built by other, more industrious bird species. Any nest large enough and in reasonable repair (because the owls cannot or will not renovate it themselves!) will do – most often the reused nest originally belonged to a large corvid such as a Carrion Crow, or a large raptor like a Common Buzzard.

Ground force

Short-eared Owls usually nest on the ground, picking a hollow (which they may deepen by scraping it out some more) that has some kind of shelter overhead. Long-eared Owls also occasionally nest on the ground. Little Owls, which are hole nesters, sometimes use old animal burrows. This is a behaviour that their American cousin, the Burrowing Owl, has taken to its extreme – this species always nests in burrows left by prairie dogs and similar tunnelling mammals and spends almost all its time on the ground. Not surprisingly, it is a champion sprinter among owls.

Above: This nest probably once belonged to a crow but is now home for a Long-eared Owl family.

Below: Short-eared Owls usually nest on the ground, among sheltering vegetation.

Eggs and incubation

Right: A clutch of Long-eared Owl eggs.

Owl eggs are white – the most common colour for hole-nesting birds, as camouflage isn't needed. They are also unusually round, compared to the familiar oval shape of a chicken egg. The clutch size is usually between three and five, with clutches tending to be larger further north in the species' range. Short-eared Owls may lay eight or more eggs in a clutch if it happens to be a 'boom' vole year. In years where food is desperately scarce, though, pairs may 'decide' not to try to breed at all. The BTO's Nest Record Scheme has given the following data for average clutch size and incubation duration:

Barn Owl	4–6 eggs	33 days
Little Owl	3–5 eggs	28 days
Tawny Owl	2–4 eggs	30 days
Short-eared Owl	4–8 eggs	26 days
Long-eared Owl	4–5 eggs	28 days

The female does all the incubation, while her mate brings food for her. She usually lays one egg each day, or every two days in larger species, and starts to incubate after the arrival of egg one or egg two. This means that by the time the last egg is laid, the first egg

has a considerable developmental 'head-start', and the eggs will hatch at different times. This is in contrast to many other birds, which only begin incubation when the clutch is complete – their eggs then all hatch more or less together, and this has the advantage of reducing the total time that there are vulnerable chicks in the nest.

Asynchronous hatching has its advantages as well. The oldest chick may be a week old or more before the youngest has even hatched, and when food arrives it easily beats its younger siblings in the scrum to be fed. In a good season with abundant prey, eventually, all the chicks will be fed, but if food is scarce, the youngest will consistently miss out and will probably soon die. Perhaps the next youngest will as well – and so on until the brood is reduced to a size that the available food can support. In very bad years it could be only the oldest chick that makes it. But it will be a well-fed and healthy chick with a good chance of surviving to breeding age. This is a better bet for the parents than rearing more but less healthy chicks.

Above: This female Short-eared Owl could be sitting on 10 eggs or even more.

Below: The youngest of this brood of seven Barn Owl chicks is barely half the size of its oldest sibling.

Life in the nest

A chick breaks out of the shell using a hard 'button' on the tip of its otherwise still rubbery bill – this 'egg tooth' gradually disappears after the first few days of life. The hatchling is feeble, with sparse pale down and closed eyes. It cannot regulate its body temperature so needs to be brooded by its mother. It can, though, raise its head and squeak for food.

Above: In this large Barn Owl brood, the smallest chick is at risk of becoming an extra meal for its big brothers and sisters.

Over the next few days, the male brings prey to the nest and the female rips it up and carefully feeds pieces to the chicks, attending to the loudest squeaker first. The owlets develop rapidly – though larger species take longer to reach fledging age than smaller ones. After the first week or so they can stand up and the eyes are beginning to open. Their mother starts to feed them whole prey items, and they begin to produce pellets. At two weeks old the body down is being replaced by 'mesoptile' plumage, a very short-term set of soft and downy true feathers which bear a camouflaged pattern, and the flight and tail feathers are growing well. At three weeks old they can walk about and stretch their wings, the eyes are fully open and in some cases the owlets are starting to think about leaving the nest.

Branching out

A nest offers safety and shelter, but as the chicks grow bigger they need to be brought food by both parents. Without mum around full-time to see off any threats, a nest can become more of a trap than a shelter, and so owlets often leave the nest well before they are able to fly, and spread out in the nearby area, reducing each individual's chance of being discovered by a predator.

This is easiest for ground-nesting Short-eared Owls, whose chicks may disperse from the nest at as early as two weeks. For Tawny Owls nesting in tree holes, the chicks climb out at four to five weeks old and scramble to nearby branches. They will not be able to fly for another couple of weeks, but if they fall to the ground they can climb back up into the tree. Only where there is no safe way to move away from the nest on foot (which means that the most dangerous predators cannot reach the nest anyway) do the young stay in or very near the nest until they are old enough for their first flight. This is often the case with Barn Owl nests, where all the chicks can do is shuffle a short distance along a ledge.

Above: A Little Owl chick, a few weeks out of the nest, works on its ground-hunting skills.

Below: Tawny owlets can climb well and perch safely on branches long before they can fly.

Right: A Long-eared Owl in mesoptile plumage has a much fluffier look than the adult bird.

The chicks continue to need parental support for some weeks after leaving the nest. It will take several weeks between their first flight and their mastering the art of hunting enough that they can live independently. In the meantime, they stay in their hiding places, only giving themselves away when a parent arrives with food. Hearing the 'squeaky-gate' begging calls of young Long-eared Owls is often the first sign that this secretive owl has a nest nearby. They take a keen interest in other living things from an early age and will snap at nearby insects. Serious hunting attempts, though, will only begin when they are flying well and the mesoptile plumage is replaced with the first set of adult-like feathers.

Right: Even at this tender age, a Long-eared Owlet can still produce a convincing wings-up threat display.

How long young birds stay in their parents' territories varies from species to species. The longer the better for their own survival, as they are building their skills in a relatively rich and safe environment, so it is in the parents' interests to allow this for a time. But as autumn advances and hunting becomes more difficult, and the young birds begin to feel the need for a territory of their own, the parents become less tolerant. Barn, Tawny and Little Owls may still be in their parents' territory as late as October or November, but juvenile eared owls move on earlier (as do their parents). They move in random directions away from 'home', with the two migratory species then starting to head south and towards coastal areas.

Above: A fledgling Tawny Owl looks very appealing, but should be admired from a distance for its own wellbeing, and yours as well because its protective parents are almost certainly nearby.

Life and Death

Once a young owl moves away from its parents' territory, it is on its own, and in a way it will always be on its own. With luck, it will find a mate, and that will allow it to raise a family, but it will be responsible for looking after its own basic needs. Experience is key to this. The owl faces an extremely steep learning curve in the art of survival through its first few months, but if it gets to a year old, its chances of reaching two years old are much improved. But every day and night brings hazards and, sooner or later, one of those will get the better of the owl.

Predation

Owls are predators themselves, but that is no deterrent at all to other, bigger predators – and indeed might even be an incentive (see intraguild predation, page 61). Birds of prey are the biggest danger. Two species in particular, the Peregrine Falcon and the Goshawk, are capable of killing any of our owls. In some parts of upland Scotland, Golden Eagles are also a danger to Short-eared Owls. As our smallest owl, the Little Owl may also be taken by Sparrowhawks. Tawny Owls are only really likely to fall prey to Goshawks, as Peregrines avoid woodland.

Mammalian predators pose little risk to healthy adult owls going about their daily business. For youngsters, though, and for incubating females, the risk is higher. The small mustelids (members of the weasel family) can climb quite well and may be able to access owl nests. Under normal circumstances, a female Tawny Owl would have little to fear from a Weasel or even a Stoat but if one entered the nest without her realising, then she would struggle to fight it in the restricted space of the nest cavity. The Pine Marten, a much larger mustelid and a very adept climber, is a serious danger to woodland-nesting owls, though is rare over most of Britain.

Above: Pine Martens in northern Europe can be significant predators of owl nests.

Opposite: Making it through winter can be a real challenge for all of our owl species.

Nest raiders

Above: Foxes are eager eaters of eggs and chicks, but they are not expert climbers so most owl nests are out of their reach.

The female owl's role in breeding is to defend the nest, and so she rarely leaves it unattended until the chicks are quite well grown. But even a short absence may be enough for a sneaky thief to make its move. Many different animals will take eggs or nestlings if they get the chance, including Magpies, Jays, squirrels (Red and Grey), mustelids, some birds of prey (including other owls), Foxes, and even Badgers and Hedgehogs if the nest is low enough to be within their reach. Owls will not typically try to nest again that year if they lose a clutch or a brood, and the following season are likely to try to nest in a more secure location.

Right: Opportunist foragers, Hedgehogs will certainly not spurn the chance to raid a bird's nest.

Left: A Tawny owlet that falls to the woodland floor needs to climb back up to the safety of the trees quickly.

Young owlets that have left the nest, but are still small and flightless, are also very vulnerable. Keeping quiet and hidden and, if possible, out of reach is their best chance of survival. Sparrowhawks and Goshawks are the main threat to the woodland-nesting species, while Foxes, mustelids and corvids are dangers to those on the ground. Mortality in the period between fledging and independence is usually higher than in-nest mortality.

Below left: Peregrines are powerful open-country hunters, more than capable of killing Barn and Short-eared Owls.

Below right: Though it is rare in Britain, the Goshawk is a real danger to woodland owls like Tawnies and Long-eareds.

Other dangers in the nest

Above: Owl nests are usually enclosed and fill up with droppings and prey remains, making them less than sanitary.

Owl nests, particularly those in holes, are not the most hygienic places, with droppings, pellets and prey remains all encouraging extra little life forms. There is a risk that the female and chicks could catch a disease or acquire a heavy load of parasites that could weaken or even kill them. In North America, the Eastern Screech-owl has the fascinating habit of keeping a 'caretaker' in its nest to help clean up. It releases a live blind snake into the nest, and this creature eats up flies, maggots and other undesirables.

In Britain, all active birds' nests are protected by law and cannot be destroyed, damaged or moved. However, this doesn't stop nests being destroyed (accidentally or deliberately) when trees are felled or buildings are demolished, which is why it's so important that any work of this kind where there might be nesting birds is done outside of the breeding season. Nesting trees may also collapse naturally, as happened in 2015 to the well-known and very long-established Tawny Owl nest tree in

Kensington Gardens, Central London (luckily not during the breeding season). Long-eared Owls nesting in old nests of other birds may lose their nests in high winds, and ground nests of Short-eared Owls could be flooded out. Barn Owls and Little Owls nesting in buildings could have their access blocked halfway through the season if the building's owner is not aware of the owls' presence and begins renovation work.

Above: Nesting in holes in big, old, partly decayed trees has one obvious drawback.

Below: A nest hole that suits Tawny Owls will also suit Jackdaws, which can lead to conflict – dangerous for both species.

A very unusual cause of death of nesting female Tawny Owls has been recorded a number of times – when a Jackdaw pair built their own nest in the same hollow as a nesting Tawny, literally on top of her as she incubated, trapping and starving her. Why these owls allowed this to happen is a mystery – somehow the Jackdaws didn't trigger the usual nest-defence reaction.

Accident and injury

Right: Barn Owls are frequent casualties on country roads because of their flight and hunting habits.

Death by misadventure is a common epitaph for adult owls, though less so for the most sedentary species which know their territories inside out and rarely venture into areas unknown. Aerial hunters are more at risk than those that usually hunt from a perch, and Barn Owls seem particularly accident-prone. Their habitat of lowland countryside, often near habitation, and their habit of flying along hedgerows to try to flush out roosting birds means that they are often hit by cars. They also often drown when trying to drink from deceptively deep water tanks on farmland. A much-publicised photo of a dead Barn Owl caught in the metal frame of a Chinese lantern has been used to try to discourage people from releasing these objects into the sky. How this accident happened isn't clear but a hunting Barn Owl frequently drops feet first into long vegetation without necessarily seeing what might be hidden in the grass. Collisions with objects like power lines are not such a serious threat to owls as they are to faster-flying birds but accidents can still happen, particularly if a nocturnal owl is forced to fly in daylight.

Below: This Eastern Screech-owl has lost an eye, which will impede its hunting ability.

Rescue and rehabilitation

Owls that have been hurt but not killed will be easy targets for predators, but if they are lucky, a kind member of the public will find them first and take them to a wildlife rehabilitator. If you happen to find an owl that is obviously injured and in need of help, you should be able to catch it by gently dropping a towel, coat or something similar over it and then gently bundling it up. Be very careful of its talons – it is much more likely to try to hurt you with these than with its bill.

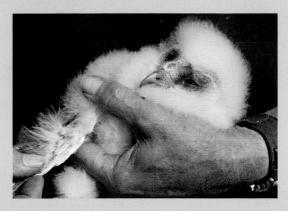

The website helpwildlife.co.uk has practical advice and a lengthy list of wildlife rescue centres. Or try your local vet, who should have contacts with wildlife rehabilitators in the area. Given prompt treatment, the owl has a good chance of survival. Those with wing injuries may never recover enough of their flight ability to be releasable but can live happily enough in captivity, and form close and rewarding bonds with their keepers. Such birds can be effective ambassadors for owl-kInd, allowing people the chance to see them up close and learn about what we can do to safeguard our wild owls.

Above: When nests are inadvertently destroyed, it may be necessary for wildlife rehabbers to hand-rear owlets.

Below: Wing injuries need very careful treatment if they are not to spell the end of an owl's hunting career.

Starvation

Above: Provisioning a growing brood is hard work, and if prey is scarce then the parents will not be able to keep all the chicks alive.

Dying from a lack of food is probably the main cause of death for owls, particularly young owls. This is exactly what is expected for a high-level predator and is the reason why it is close to impossible for predator populations to outstrip those of their prey species in most ecosystems. Some chicks starve in the nest while still very small – the youngest in the brood are most at risk, as their older siblings commandeer the lion's share of the food – and this continues until the youngsters are hunting for themselves. Now, they have to develop good hunting skills and fast, because winter is coming. For all owls, food is harder to find in the colder months – and the risks of going hungry are higher when temperatures are lower. For territorial species, finding a place where they can hunt consistently well without being seen off by other owls is crucial. Territory-less young Tawnies have to take far more additional risks – moving around more, perhaps hunting in the daytime, perhaps checking roads for carrion – so are more vulnerable to other dangers as well as to starvation.

When numerous Short-eared and Long-eared Owls arrive in Britain in winter, this is a treat for birdwatchers, but it is hunger that has driven these birds here and many will starve. A majority will be youngsters, which have less refined hunting skills than adults and may also be weakened by long flights powered by not enough food. Emaciated, weakened owls are sometimes found helpless but alive, and may be saved by rehabilitators, but this level of hunger has often already done irreversible damage to their organs.

Above: Hungry Short-eared Owls arriving on Britain's coastlines in autumn will gather at good hunting grounds.

Disease and sickness

Above: Roosting in groups offers more safety from predators for Long-eared Owls, but may make it easier for disease to spread.

Because of their solitary ways, most owls are at relatively low risk of contracting infectious illnesses directly from other owls – the communal roosts of Short-eared and Long-eared Owls being the exception. There are other ways they can catch diseases, however. Avian malaria, transmitted by mosquito bites, has been noted with increasing frequency in owl populations, the rise being linked by some to climate

Trouble at the top

One of the problems with being at or near the top of a food chain is the accumulation of poisons that were ingested by prey. The use of DDT and other persistent toxins to kill crop pests has had well-documented disastrous effects on predators, including owls. The decline of Barn Owls across Britain in the 1950s to 1970s was partly due to the use of these chemicals, which killed and weakened birds and caused them to

Above: Powerful and environmentally persistent pesticides like DDT can seriously harm the breeding success of Barn Owls.

lay thin-shelled, non-viable eggs. DDT use was banned in Britain in 1986, and this has helped Barn Owl numbers to recover. However, it and other persistent poisons can still be found in owl bodies because of its extremely slow rate of breakdown in the environment.

change. Some birds seem to be symptomless carriers, but others become ill and die (though the infection is treatable if the bird is rescued in time). Avian forms of cholera and tuberculosis have also been noted in wild owls. Almost all owls carry parasites which feed on feathers and blood, without harming the bird, but occasionally the parasite load can become heavy enough to cause problems.

Persecution

Deliberate killing of owls by humans was once commonplace, as owls (like all predators) were viewed as unwanted competitors for game like Pheasants and Rabbits. They were killed by poisoning, shooting and trapping, with the brutal leg-holding gin trap, positioned on top of a post, a particular favourite. Today, owls and all other birds of prey are fully protected from this kind of persecution. However, illegal persecution continues. Owls are not targeted with anything like the same frequency as certain day-flying raptors like Hen Harriers, Peregrines and Common Buzzards, but there are still incidents each year. The RSPB publishes an annual report, *Birdcrime*, listing persecution incidents. The most recent report (at the time of writing) covered 2014 and listed shootings of three Barn Owls, one Long-eared Owl, one Little Owl and two Tawny Owls, plus two Barn Owl nests were destroyed and four trapping incidents involving owls were reported.

Right: Gin-traps have long been illegal in Britain, but there are still incidents of their use.

Beating the odds

With good luck and good survival skills, owls can live relatively long lives. In captivity, many species can reach their 20s or 30s, which shows the potential lifespan in the absence of most of the hazards that face wild birds.

Recoveries of ringed birds are the source for longevity records in the wild, and the ringing schemes for various countries have been underway for long enough now that we can be reasonably confident that they are truly representative of natural lifespans in most cases. In Britain, the record holders are as follows:

Above: This Long-eared Owl chick now carries a unique identifying leg ring, so its life history can be traced if it is ever refound.

Barn Owl	15 years 3 months
Tawny Owl	21 years 10 months
Little Owl	10 years 11 months
Long-eared Owl	12 years 10 months
Short-eared Owl	6 years 7 months

For some species, European ringing recoveries have produced longevity records that approach or exceed these – quite dramatically so in the case of Short-eared Owl.

Barn Owl	17 years 11 months (the Netherlands)
Tawny Owl	22 years 5 months (Czech Republic and another in Finland)
Little Owl	10 years 2 months (Denmark)
Long-eared Owl	17 years 11 months (Finland)
Short-eared Owl	20 years 9 months (Germany)

Below: Ringing has taught us much about bird migrations, survival and longevity.

So a few wild owls do make it to a grand old age. They are most unlikely to die peacefully, though, but will probably succumb to starvation as soon as their hunting skills start to deteriorate, especially if driven from their territory by a younger and stronger bird. However, with all those years under their belts they are likely to leave behind a good number of offspring.

Owl Conservation

The International Union for Conservation of Nature and Natural Resources (IUCN), an international organisation concerned with nature conservation, produces a Red List of species that are of conservation concern. It assesses all five British owl species as 'Least Concern' – meaning there is no immediate danger of their becoming extinct. This is not surprising as all have extensive distributions and large populations. However, two species (Long-eared and Short-eared) are thought to be declining on a world scale and continued declines will eventually lead to their being 'upgraded' to one of the categories that denote high conservation concern.

Global and local

The IUCN's categories range from Least Concern through Near Threatened to Vulnerable, Endangered and Critically Endangered, indicating increased severity of threat and likelihood of extinction. The final two categories are Extinct in the Wild and Extinct. Worldwide, about 75 owl species fall into the threatened categories. Some of those classed as Critically Endangered may already be gone – the Pernambuco Pygmy-owl, for example, has not been seen in the wild since 2004. Other seriously threatened species may be pulled back from the brink – considerable efforts are underway to save the spectacular Blakiston's Fish-owl of East Asia, one of the world's largest owls.

Although our five owls are not in trouble (yet) globally, two are of conservation concern in Britain. The RSPB classes the level of conservation concern for each British species by a traffic-light system – green, red and amber. The Short-eared Owl is amber-listed, and while the Barn Owl has now changed from amber to green, the Tawny Owl has gone the other way and has been amber-listed since 2015. The Long-eared Owl is green-listed, and the Little Owl, as an introduced non-native species, is not given a category.

Above: Blakiston's Fish-owl is one of the world's largest – and most threatened – owls.

Opposite: Although it is not common, the Long-eared Owl is not considered to be threatened in Britain.

A thought for the little ones

Above: A favourite of birders, the Little Owl is in decline in Britain.

Many British birdwatchers are surprised to learn that the Little Owl is not a native species. It is rather an oddity among the various (deliberately or accidentally) introduced birds that have made themselves at home in the wild here in Britain. Some have caused obvious problems – one or two have even been culled on the grounds that they could pose a serious risk to human and/or conservation interests. Many others (most notably the Ring-necked Parakeet with its outrageous neon-green plumage and deafening shrieks) seem very out of place here. But the Little Owl appears to have fitted in beautifully – there are very few obvious conflicts with native species and certainly no one is advocating that it should be eradicated. This is largely down to it being a widespread bird on mainland Europe, where it has long coexisted happily with many of the same species that live in Britain. Its great charm makes it a favourite with birdwatchers, as well.

Like other non-native birds, it is not a conservation priority for the RSPB, as resources must be directed towards species that occur here naturally. However, it enjoys the same blanket protections as all other wild birds in Britain. Going by BTO data, it may be declining at present, as are many farmland and open-country birds. However, the RSPB's work with land managers to improve such habitats for all wildlife will benefit this species as well.

Above: Ring-necked Parakeets and Little Owls, both non-native in Britain, may compete for nest sites.

Left: Heather and grassy moorland is breeding habitat for Short-eared Owls in Britain.

103

CONSERVATION

The amber-listing of the Short-eared Owl reflects declines in the (already quite small) British breeding population, a situation reflected in mainland Europe. These declines do follow a period of increase, probably as swathes of upland were planted up with conifers. The young plantations suited nesting Short-eared Owls well, but as the trees grew and shaded out the grassy understorey, the owls could no longer breed in such areas.

Below: Undisturbed open moorland, breeding habitat for Short-eared Owls, is scarce in Britain.

Above: Dense pine plantations offer little in the way of nest sites for owls.

Below: Although it remains our commonest owl, the Tawny is declining at a worrying rate.

The Tawny Owl's elevation to amber status follows a steady decline, as revealed through a range of BTO bird surveys from Garden BirdWatch to the Nest Record Scheme. This consistent long-term trend has seen the Tawny population in Britain fall by a third over the last 25 years. Why the species is declining is not yet known – many factors could be at play, from loss of suitable nest sites or hunting habitat to declines in important prey species. While the extent of woodland cover in Britain has risen from 5 per cent of land area at the start of the 20th century to 13 per cent today, other factors such as type of woodland, and average distance between woodland patches, are also important. Gardens are important habitats for Tawnies as well, but trends to pave over lawns, tidying up 'wild' areas that harbour small rodents, and the felling of mature native trees all make gardens a lot less Tawny-friendly.

Barn Owl – bouncing back

Above: The prosperous late 20th century meant many old farm buildings were 'done up' and rendered Barn Owl-unfriendly.

Barn Owl numbers took a serious hammering through much of the 20th century, thanks to the triple whammy of farming intensification, use of deadly DDT and other pesticides, and widespread renovation of derelict farm buildings. Conservationists and bird-lovers were alarmed by the wholesale loss of one of our most iconic and beautiful owl species and some responded by releasing captive-bred Barn Owls into the wild. Ad hoc, unplanned releases like these, without any attempt to prepare the birds for a life in the wild, are a very bad idea. With the factors causing the decline still very much there, the released birds were even less likely to survive than their wild-born relatives. Such releases are now illegal.

Over the last couple of decades, things have looked up for Barn Owls. Habitat improvements, a ban on DDT, and more and more landowners providing nest boxes are all helping to address the problems. A run of mild winters in the 1990s also helped to improve survival, and Barn Owls were clearly doing much better by the 2000s. They took a further knock around 2009–2011, when severe winter weather took its toll, but overall were considered to be doing well enough to be moved from the amber list to the green list in 2015.

Above: By and large, farmers are delighted to see this beautiful owl hunting on their land.

How you can help

There are many things that you can do to help Britain's owls to thrive. The key thing to remember is that owls sit at or near the top of a pyramid of life – without the 'lower' animals that are their prey and the invertebrates and plants that the prey species eat, there can be no owls. So conservation really does have to start at the grass roots.

Habitat support

If you have a garden, you may have visiting Tawny Owls. If your garden is big with tall trees, you could even have breeding Tawnies. In any case, your garden is part of a network of green space in your local area, and anything you do to improve its attractiveness to wildlife will help improve the whole of that green space. Simple measures like leaving patches to go wild, planting native rather than non-native plants, and keeping pesticide use to a minimum will all help.

If you do have Tawnies locally (you will know by the hooting in late winter and early spring if you do) and you have a suitable big tree, you could try putting up a nest box for them. However, it's very important that nest boxes for Tawny Owls are positioned well away from houses or other places with frequent 'human traffic', as these owls can be so aggressive and potentially dangerous in the nesting season. If you already have Tawnies nesting near your house, give the birds plenty of space and make sure that children, in particular, stay well away.

Further afield, you can help support good wildlife habitats by buying organic food where possible, by supporting conservation bodies and their nature reserves through donations, volunteering or both, and by keeping an eye on local developments and speaking up for wildlife when green spaces are under threat. If you have seen Barn or Little Owls locally you could contact the landowner and see if they might consider putting up a nest box. And if you have a wildlife rehabilitation centre nearby you can support them through donations and perhaps also volunteering – few things could be more directly rewarding than helping an owl to recover from mishap and return to the wild.

Opposite: It may take a while for a pair of Tawnies to discover a new nest box, but once they do they are likely to use it for years.

Worldwide threats

Because loss of habitat is the main threat to wildlife, species can best protect themselves by having a wide distribution and the adaptability to use varied habitats. The five British owls are generally equipped to cope, but many other owl species in the world are much more limited and therefore much more vulnerable. Those that are confined to just one or a few particular islands are particularly at risk. For example, many of the Philippine islands have their own unique species of scops-owl, and nearly all of these are declining because of deforestation, with little hope of any improvement in their situation. The wild landscapes on which owls depend have been dangerously diminished as expanding human populations have driven the development of farmland and cities.

Climate change is another growing issue. Its most obvious potential victims are the most northerly-breeding owls, such as Snowies, but species that live in

Below: The Barred Owl of North America prefers patchy woodland. Forest clearance is allowing it to spread westwards, where it is displacing the endangered Spotted Owl.

mountainous areas in temperate and even tropical regions could also 'run out of room', and increased drought and flooding could impact many others.

Sadly, some owls are still persecuted deliberately – fishermen in East Asia trap and kill the mighty Blakiston's Fish-owl as it competes with them for the fish that are their livelihood. Meanwhile, some cultures, for example in Jamaica and Madagascar, hold strong anti-owl superstitions and deliberate killings for this reason are frequent. Owls are also hunted for bushmeat in parts of Africa.

Above: Snowy Owls depend on high cold tundra to breed, and their current decline may be linked to climate change.

Left: Serendib Scops-owl, native to rainforest in Sri Lanka, was discovered in 2001, and surveys since then suggest it has a population of just 200–250 individuals.

Making and siting a nest box

The RSPB sells nest boxes suitable for Barn and Tawny Owls - visit: http://shopping.rspb.org.uk/birds-wildlife/nestboxes/bird-of-prey-nestboxes.html If you would like to make your own Barn Owl box, search the RSPB website for 'barn owl box' to find articles with full instructions. For building other kinds of boxes, buy the *BTO Nestbox Guide* for comprehensive guidance.

Tawny and Little Owls will appreciate some sheltering vegetation around the box, so it is not obviously in view to all and sundry, while Barn Owls will use boxes in more open situations. Nest boxes should always be placed out of direct sun and out of easy reach of predators, including curious humans, at least 3 metres (10 feet) above ground height.

BELOW: Nest boxes should be mounted well above ground level, and out of direct sunlight.

Watch owls, respectfully

Most people love owls a little (or a lot) more than they love other birds – owls certainly have a special place in our consciousness. Those of us who see owls regularly cherish those encounters and those who don't will usually jump at the chance to see them if they can. Watching and enjoying owls is a great way to build concern and awareness about these charismatic birds, and encourage people to become invested in wildlife conservation.

It is very important to remember that these birds are sensitive to disturbance and must be treated with care and respect at all times. If you find a nest, watch from a distance and be careful who you tell about its location – it's better to keep things quite vague, and the same goes for roosting owls. Birdwatchers and photographers who try to get just that little bit closer can cause an owl to abandon its nest or roost. The jury is still out on whether artificial lights upset owls, but it is best to err on the side of caution and avoid flash photography or shining lights onto owls at night.

Above: When Short-eared Owls gather to hunt in winter, birders and photographers can enjoy wonderful views.

Below: Long-eared Owls will use the same roost spots all through winter if they are undisturbed.

Survey the scene

There are numerous wild bird surveying projects in Britain to which you can contribute. Most are organised by the BTO, and they range from Garden BirdWatch (where you submit a list of sightings In your garden week on week) to the Breeding Bird Survey (where you walk a designated transect route through a 1km square (0.38 square miles) twice during the breeding season and record all bird-breeding activity). The RSPB also runs surveys, such as the Big Garden Birdwatch every January. If you want to get really involved, you can train as a bird ringer and work on the 'front line' of wild bird surveying.

BELOW: Training to be a ringer is a great way to work on the 'front line' of wild bird research.

Watching hunting Barn and Short-eared Owls is the easiest way to enjoy owl encounters, as both species will hunt in daylight and both hunt on the wing. Avoid chasing the owl, and allow it to come to you. Hunting birds use the breeze for uplift, so position yourself with the wind behind you and the owl will, with luck, move towards you. There are nature reserves up and down the country that are great for owls. You can find out more about some of these reserves here: www.rspb.org.uk/discoverandenjoynature/seenature/reserves/tags.aspx?tag=owls

Opposite: Watching from a permanent hide is the best way to enjoy close views of birds with no risk of disturbance.

Owls in Culture

Owls inspire and mystify us, and they always have. They conduct their business under cover of darkness, they are more often heard than seen, and some of the things they can do seem close to magical. There are enough folk tales and legends about owls to fill another book, and many of them cast the owl as a figure to be admired, respected and perhaps also feared.

Myth and folklore

Across many cultures, the owl (or more specifically its call) foretells bad news of some kind, often death. It is said that calling owls foretold the deaths of several Roman emperors. In Britain, the Tawny's hoot is the most often-heard and distinctive owl sound, and it has been associated with many things, from announcing the loss of a young woman's virginity, to predicting the sex of an unborn child (if you hear a hoot while pregnant, your child will be a girl). The poet John Ruskin said that its call foretold unspecified 'mischief', and the appearance of a hooting and shrieking 'bird of night' in Shakespeare's *Julius Caesar* was one of several early portents of the trouble that was to follow.

Above: The hooting of a Tawny Owl is often considered a portent – sometimes good, sometimes bad.

Left: These silent night flyers both inspire and unnerve us.

Opposite: The expressive face of a Long-eared Owl speaks volumes, but who knows what the bird is really thinking (if anything).

Little and wise

Above: Ancient Greek coins show the goddess Athene, and her owl alter ego.

The Little Owl's short tenure as a British bird means that there is little home-grown folklore around it, but it is one of the most significant animals of all in ancient Greek stories, as the icon of the goddess Athene (and its genus still bears her name). Athene was associated with wisdom, and she was either shown with an owl or the image of an owl was used to represent her on all kinds of decorative works. Today the owl image can still be found on the Greek 1 Euro coin. The Little Owl is the likeliest candidate to be the species the Greeks had in mind, going by the images of a smallish, big-eyed and flat-headed owl, and the fact that Little Owls are common around the Mediterranean and often live near habitation.

As well as wisdom, Athene's remit included inspiring warriors to victory. At the Battle of Salamis, it is said that the Greeks were spurred on to defeat the Persians, despite being outnumbered, when an owl came and settled on the mast of Admiral Themistocles' flagship. The Greek soldiers believed that this was a show of support from Athene and were inspired to a famous victory.

Left: A Little Owl in modern Greece, sheltering at one of the Meteora monasteries in Thessaly.

Left: Owl eggs and owls themselves were used in folk medicine to treat various maladies.

Owls had their uses in folk remedies and superstitions. A broth made from owls was said to cure whooping cough, and a dead owl nailed to a barn door was a popular trick to ward off lightning strikes and other bad luck until well into the 19th century.

Below: Blodeuwedd, the beautiful but devious Welsh anti-hero of legend, before she was transformed into an owl.

An alcoholic could be cured by eating raw owl eggs, and if a child partook of the same meal then he or she would never become a drunkard. Another use for owl eggs was to improve vision, but in this case they had to be burned to charcoal and then the ashes made into a potion.

The Barn Owl's chilling scream gave rise to its alternative name of 'screech-owl' and that, along with its ghostly appearance, inspired a general belief that it was a bird of evil omen, foretelling death, or at least a spell of stormy weather. The Welsh story of the almost immortal hero Lleu Llaw Gyffes tells of how his wife Blodeuwedd ('flower face') betrayed and tricked him into revealing how he could be killed. She murdered him and stole his lands, but the magician Gwydion punished her by turning

ALAN GARNER

THE OWL SERVICE

"Alan Garner's fiction is something special." Neil Gaiman

her into an owl – usually stylised as a Barn Owl in pictorial representations, although the story does not specify a species. Her tale is told in the book *The Owl Service*, by Alan Garner, a story for teenagers and young adults.

The Inuit bring us a story of how the owl got its distinctive face shape – the story may have been inspired by Snowy Owls but possibly also Short-eared Owls. The first owl was originally a little girl but a witch turned her into a bird. In a panic, and underestimating the power of her new wings, the bird flew into the side of a house and squashed her bill into a hook and flattened her face, becoming an owl.

Left: The story of Blodeuwedd is beautifully adapted for teenage readers in *The Owl Service.*

Below: Did the Snowy Owl get its hooked bill after an unfortunate collision with a wall?

Children's stories

Owls appeal to us universally, but perhaps to children most of all. From babyhood we humans are programmed to gaze at anything that looks like a pair of eyes in a roundish face, and the human-like qualities of the owl face seem well designed to trigger that instinctive response. It's no wonder that owls are such popular characters in children's stories, both in print and on screen. These fictional owls are often admirable characters as well, lacking the darker elements that characterise owls in folklore and legends, though they can be prone to grumpiness, pomposity and often have the general air of the schoolmaster.

The expressive owl face, with its flat profile and forward-facing eyes, strikes us as intelligent (or at least more intelligent than the beaky visage of the typical bird, with its side-mounted eyes that can only look at us one at a time). Many owls in stories are there to provide a voice of mature wisdom. Disney's 1981 animation *The Fox and the Hound*, with its

Below: Big Mama, the wise and motherly owl who took care of Disney's 'Fox and Hound' when they were still 'Cub and Puppy'.

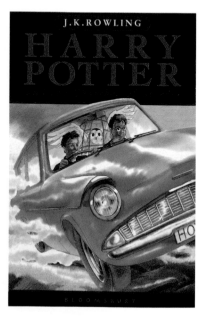

cast of woodland and domestic animals, featured Big Mama, a particularly cuddly owl who was as caringly maternal as she was wise.

Another Disney owl began his life in print. Winnie-the-Pooh's entourage included the owl whose name was Owl (but he spelled it 'Wol') – an avuncular bird whose slightly patronising and pompous manner earned him the respect of the rest of the gang, but who was in fact nowhere near as wise or learned as he liked to think he was. In Disney's film adaptations of the stories, Owl was a North American Great Horned Owl, but the author of the Pooh stories, A. A. Milne, was an Englishman who set his stories in Ashdown Forest, Sussex, and his Owl (as illustrated by E. H. Shepard) was unmistakeably and appropriately a Tawny.

Harry Potter's adventures in the world of wizardry were also replete with owls of various species. The owls are messengers, conveying letters and sometimes other things between magical folks, and are also kept as much-loved pets. Harry's own owl, Hedwig, was a Snowy Owl, but many other species appear in both the books and the films.

Owls in falconry

Using birds of prey to hunt game is a practice that has been carried out for millennia, but in Britain had its heyday in the Middle Ages. With the advent of the firearm, using birds of prey to hunt fell from favour, and soon the wild predators were being killed off as competitors. Today, falconry has enjoyed something of a revival as a leisure sport and is subject to various regulations – most notably that it is no longer legal to take birds from the wild to keep and train for falconry.

Owls have traditionally been quite unpopular as raptors for falconry, as most are nocturnal and in general the prey they prefer is too small to be of interest to falconers. They are not regarded as particularly trainable either – but the most 'easy' falconry birds (the true falcons, and the Harris Hawk of North America) are those that sometimes naturally hunt in pairs or groups in the wild, so are hard-wired to be cooperative and responsive to others. In modern falconry, owls are rather

Above: Owls often feature in crowd-pleasing falconry demonstrations, especially the large eagle-owl species.

Below: For practical falconry, raptors are favoured over owls.

Right: The Eurasian Eagle-owls living and breeding in Britain today may all descend from escaped falconry birds.

Opposite: Falconers build great understanding with their birds, but this is only possible through knowledge of the owls' nature as wild creatures.

more popular, as much for their beauty as anything else. In commercial shows, trained birds are often used to demonstrate natural hunting behaviours – so a Little Owl, for example, might be trained to run after a lure on the ground, or turn over cups to find a treat.

One of the big problems with owls being used in falconry is that they sometimes escape. Many escapees are recovered sooner or later, and the increasing use of telemetry devices to track them down is helping with this. But escaped owls can cause havoc, particularly the large species such as the Eurasian Eagle-owl and Bengal Eagle-owl – the latter is particularly common and popular among falconers. Both of these birds are capable of killing pet cats and small dogs, and will also kill native birds of prey if they find them. Falconry can be a great way to learn about bird behaviour, and owls, while they may not be the most impressive hunters, do by all accounts make fascinating and delightful companions. However, it is vital for our native wild owls that everything possible is done to prevent escapes.

Glossary

Bird of prey A bird that hunts and kills vertebrate animals – usually only used for owls and raptors.

Brood The group of chicks that hatch from a clutch of eggs.

Call Any sound made by a bird that isn't its territorial/courtship song.

Clutch A set of eggs that are laid in the same nest and incubated together.

Facial disk The stiff ruff of feathers that circles an owl's face and helps to channel sound to its ear openings.

Fledge When a young bird leaves the nest.

Flight feathers The long feathers of the wing that provide lift for flight.

Genus A group of very closely related species.

Mesoptile A set of very soft downy feathers that young owls acquire before their first fully developed juvenile feathers.

Migration A regular seasonal journey after breeding, usually from a colder to a warmer climate.

Moult Shedding old feathers and growing new ones, an annual post-breeding event for most adult birds.

Nest site The place where a bird lays and incubates its eggs and rears its young – with owls usually it is some kind of hole or crevice.

Nomadism Unpredictable wanderings (as distinct from regular migrations).

Pellet A compressed mass of indigestible food remains which is regurgitated by birds of prey.

Plumage A bird's full set of feathers.

Predator Any animal that hunts and kills other animals for its food.

Preening Using the bill and feet to clean and straighten the feathers.

Prey The living animals that predators catch, kill and eat.

Primaries The outermost (longer) flight feathers.

Raptor A day-flying bird of prey, such as a hawk, falcon, kite, buzzard, eagle or harrier.

Secondaries The inner (shorter) flight feathers.

Song A particular sound made by a bird that warns rivals to stay out of its territory and (if the bird is unpaired) attracts a mate.

Species A population of birds that are all of the same 'type' and can freely interbreed.

Territory An area of habitat occupied by a bird or pair of birds, and defended from rivals.

Talons The curved claws of birds of prey.

Xygodactyl Toe arrangement seen in owls and some other birds, where two toes point forwards and two backwards.

Further Reading and Resources

Books

D. Chandler, *Barn Owl*. New Holland, London, 2011.

M. Constantine and The Sound Approach, *The Sound Approach to Birding: A Guide to Understanding Bird Sound*. The Sound Approach, Poole, 2006.

J. del Hoyo, A. Elliot and J. Sargatal, *Handbook of the Birds of the World. Vol 5, Barn-owls to Hummingbirds*. Lynx Edicions, Barcelona, 1998.

C. König, F. Weick and J-H. Becking, *Owls of the World* (Helm Identification Guides). Christopher Helm, London, 2008.

H. Mikkola, *Owls of the World: A Photographic Guide*. Christopher Helm, London, 2012.

M. Robb and The Sound Approach, *Undiscovered Owls*. The Sound Approach, Poole, 2015.

M. Taylor, *Owls*. Bloomsbury, London, 2012.

M. Taylor, *Beautiful Owls*. Ivy Press, Sussex, 2013.

D. Tipling and J. Peltomaki, *Wildlife Monographs: Owls*. Evans Mitchell Books, London, 2013.

M. Toms, *Owls* (New Naturalist 125). Collins, London, 2014.

Online

British Trust for Ornithology (studying population and distribution of birds in Britain, with details of surveys to which all birdwatchers can contribute): www.bto.org

Help Wildlife (online resource for wildlife rescue in Britain, including lists of local rehabilitators): www.helpwildlife.co.uk

IUCN Red List (global conservation categories for most species, and details of threats and conservation actions): www.iucnredlist.org

The Owl Pages (information on owl biology and conservation, also photo galleries, folklore and legend and other resources): www.owlpages.com

The Global Owl Project (bringing together contemporary research on owls worldwide): www.globalowlproject.com

RSPB (the leading body for conservation of birds and other wildlife in Britain): www.rspb.org.uk

Xeno-canto (huge database of bird songs and calls): www.xeno-canto.org

Acknowledgements

I'd like to thank Julie Bailey at Bloomsbury for commissioning this book, and Julie and Katie Read for their skill and patience in managing the project. The copy editor, Liz Drewitt, did a great job tidying up the text, and Rod Teasdale brought text and photos together to create a lovely design. Thanks also to Helen Snaith for compiling the index.

Owls are difficult photographic subjects, as I know only too well, and we are very lucky to have so many talented photographers putting in the hours to bring so many beautiful images to these pages – my thanks go to them all, and also the many fieldworkers and scientists whose research continues to give startling new insights into owl behaviour and biology every year. I'd like to thank my birding friends who've been with me on some of my most memorable owl encounters – in particular Nigel and the gang, Nick, Shane and Phil. And also a big thank you to my non-birding friends who, as always, keep me going with moral support and well-timed nights at the pub.

Image credits

Bloomsbury Publishing would like to thank the following for providing photographs and permission to reproduce copyright material.

While every effort has been made to trace and acknowledge all copyright holders, we would like to apologise for any errors or omissions and invite readers to inform us so that corrections can be made in any future editions of the book.

Key t = top; l = left; r = right; tl = top left; tcl = top centre left; tc = top centre; tcr = top centre right; tr = top right; cl = centre left; c = centre; cr = centre right; b = bottom; bl = bottom left; bcl = bottom centre left; bc = bottom centre; bcr = bottom centre right; br = bottom right AL = Alamy; FL = FLPA; G = Getty Images; NPL = Nature Picture Library; RS = RSPB Images; SH = Shutterstock

Front top, spine Mike Powles,G; **front below** Ben Queenborough,G; **Back top** Nick Cable/G; **back below** Images from BarbAnna/G; **1** Wang LiQiang/SH; **3** duangnapa_b/SH; **4** FotoRequest/ SH; **5** jack53/SH; **6** Guy Piton/G; **7** KOO/SH, bcl; **7** Piotr Krzeslak/SH, br; **9** David Evison/ SH, bl; **9** Steven David Miller/NPL, tr; **10** Gypsytwitcher/SH; **10** Szczepan Klejbuk/SH, bl; **10** Super Prin/SH, tl; **10** Brian Lasenby, br/ SH; **10** David Evison/SH, tr; **11** Luiz Antonio da Silva/SH, br; **11** critterbiz/SH, tl; **11** Ondrej Prosicky/SH, tr; **11** Don Mammoser/SH, bl; **12** Frederic Desmette/G, bl; **12** David Tipling/G, tl; **12** Andy Rouse/G, br; **13** Rolf Kopfle/G, bl; **13** Andy Rouse/G, br; **14** Randimal/SH, bl; **14** Tomas Hilger/SH, bcl; **14** Bildagentur Zoonar GmbH/ SH, tcl; **14** Tobyphotos/SH, tl; **15** Dave Chang/ SH, bc; **15** Arterra/G, tl; **15** Michal Masik/SH, tr; **16** Tony Campbell/SH; **17** Wayne Lynch/G; **18** Chris van Rijswijk/Minden Pictures/FL; **19** mycteria/SH; **20** cejen/SH, bl; **20** Nigel Dowsett/ SH, br; **20** Erni/SH, cl; **21** Ian Schofield/SH, br; **21** Adrian T Jones/SH, cr; **21** Harold Stiver/SH, bl; **22** Andrew Kandel/AL, bl; **22** Tui De Roy/NPL, tr; **23** Henk Bentlage/SH, br; **23** MarclSchauer/SH, bl; **23** Darroch Donald/AL, tr; **24** John Downer/ NPL, tl; **24** Jill Lang/SH, br; **24** Bildagentur Zoonar GmbH/SH, bl; **25** Marco Rolleman/SH; **26** Gregg Williams/SH; **27** Kay Schultz, bc; **27** Utopia_88/SH, tr; **28** DEA / G. CARFAGNA/G, br; **28** Darren Baker/SH, tl; **28** Chris van Rijswijk/

Minden Pictures/FL, tr; **29** AGF/G; **30** Fabrice Chanson/Biosphoto/FL; **31** Paul Sawer/FL, bl; **31** Mark Caunt/SH, br; **32** Stephen Dalton/NPL, bc; **32** Andy Rouse/NPL, tl; **33** Vishnevskiy Vasily/SH, tc; **33** Sharon Haeger/SH, br; **34** Milan Radisics/ NPL; **35** David Chapman/AL; **36** Tony Mills/ AL; **37** MIKE READ/NPL, br; **37** Loic Poidevin/ NPL, tc; **38** Michael Quinton/Minden Pictures/ FL, bl; **38** Andrew Swinbank/SH, tc; **39** Pascal Tordeux/NPL; **40** Michiel Vaartjes/Nature in Stock/FL; **41** Jussi Murtosaari/NPL; **42** aaltair/ SH; **43** John Waters/NPL; **44** blickwinkel/Hicken/ AL; **45** John Watkins/FL; **46** Peter Barritt/AL; **47** Graham Catley/AL, bc; **47** Frank Fichtmueller/ AL, tc; **48** Andy Rouse/NPL; **49** Loop Images/G; **50** YK/SH; **51** Dietmar Nill/NPL; **52** Ben Hall/ NPL, tc; **52** Mark Bridger/SH, bl; **53** Andy Rouse/ NPL; **54** Mark Caunt/SH, tl; **54** Edo Schmidt/AL, bc; **55** Andy Sands/NPL, tc; **55** Matt Cuda/SH, br; **56** Rolf Nussbaumer/NPL; **57** Adrian Engelbrecht/ AL; **58** Juniors Bildarchiv GmbH/AL, tc; **58** Tony Campbell/SH, br; **59** Angelo Gandolfi/NPL; **60** magicxeon/SH, cr; **60** Pete Oxford/NPL, tl; **61** WILDLIFE GmbH/AL; **62** Clement Philippe/AL, tc; **62** Sabena Jane Blackbird/AL, br; **63** John Waters/NPL; **64** Victor Tyakht/SH; **65** Jerome Murray – CC/AL; **66** Andy Rouse/NPL, br; **66** Neil Burton/SH, tc; **67** Richard Brooks/RS, bl; **67** Colin Varndell/AL, tr; **68** Jerome Murray – CC/AL; **69** Paul Sawer/FL; **70** Dietmar Nill/NPL; **71** Dietmar Nill/NPL; **72** Jose Luis GOMEZ de FRANCISCO/ NPL; **73** Loic Poidevin/NPL; **74** MZPHOTO. CZ/SH; **75** Hans Overduin/Nature in Stock/FL; **76** Harri Taavetti/FL; **77** Richard Whitcombe/ SH; Arco Images GmbH/AL; **79** Vishnevskiy Vasily/SH, tr; **79** Nature Production/NPL, b ; **80** Vishnevskiy Vasily/SH; **81** imageBROKER/AL, br; **81** ARCO/NPL, tc; **82** Photo Researchers/FL; **83** Andy Rouse/NPL, tc; **83** Menno Schaefer/SH, br; **84** yykkaa/SH, tr; **84** Diane McAllister/NPL, br; **85** Colin Seddon/NPL; **86** Tony Campbell/SH; **87** Andy Trowbridge/Nature Picture Library/AL; **88** Menno Schaefer/SH, tl; **88** Robert HENNO/AL, br; **89** Ashley Cooper/AL, tl; **89** ullstein bild/G, bl; **89** Lukas Gojda/SH, br; **90** UniversalImagesGroup/G; **91** PhotoJanski/SH, br; **91** Ortodox/SH, tc; **92** Gerrit Vyn/AL, tr; **92** Action Sports Photography/ SH, bl; **93** Toby Houlton/AL, tr; **93** Gerard Lacz/ FL, bc; **94** ullstein bild/G; **95** Roger Powell/NPL; **96** Victor Tyakht/SH; **97** Kim Taylor/NPL; **98** Roger Tidman/FL; **99** De Meester Johan/Arterra

Index